Charles F. Millspaugh

Preliminary Catalogue of the Flora of West Virginia

Charles F. Millspaugh

Preliminary Catalogue of the Flora of West Virginia

ISBN/EAN: 9783337272500

Printed in Europe, USA, Canada, Australia, Japan

Cover: Foto ©berggeist007 / pixelio.de

More available books at **www.hansebooks.com**

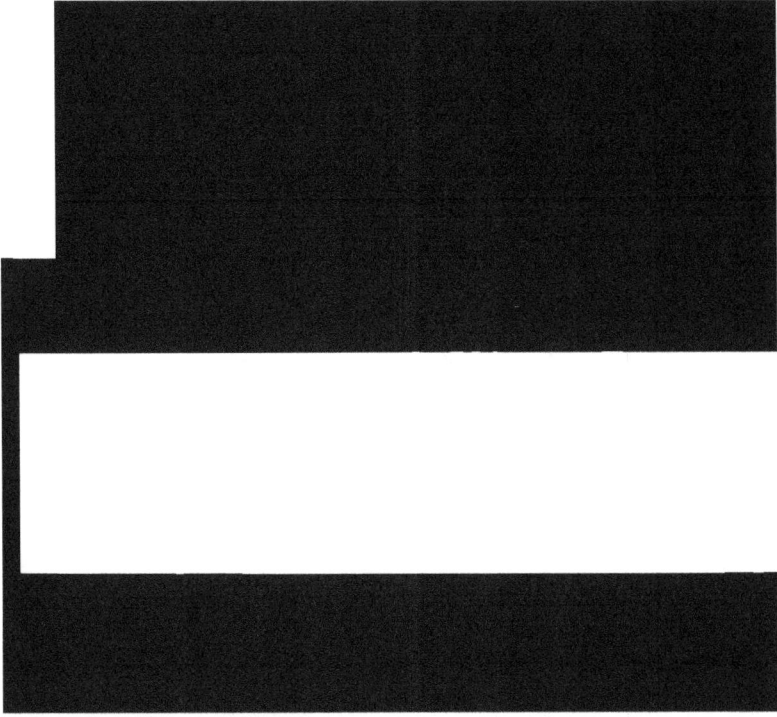

ABBREVIATIONS.

M. & G —Profs. H. N. Mertz, and G. Guttenberg.

L. W. N.—Mr. L. W. Nuttall.

V. M.— Miss Verona Mapel.

James.—Prof. Joseph F. James.

Barnes.—Prof. Charles R. Barnes.

Porter.—Prof. Thos. C. Porter.

G.—Dr. H. McS. Gamble.

W.—Dr. Roscerans Workman.

INTRODUCTORY.

In issuing this Preliminary Catalogue, I do so with no desire to convey an impression that my work in this respect is in any way complete; but simply to enter a wedge, to inspire the few who are working in the field to greater energy.

Few species are recorded in the Liverworts, and Lichens; and in the lower forms of Pond Life, and Fungi, as very little collecting or study has so far been done in these classes.

In this preliminary list I shall not enter into the details of our geology and topography, suffice it to say the State promises more of interest to the botanist than probably any other east of the Mississippi river, as it appears to be the southern limit of many boreal, the northern of many austral, and the eastern limit of many occidental forms. It bids fair also to continue to present many novelties; this may especially be said of all that unexplored and fascinating region laying south of the Great Kanawha River, a region that as far as I can learn the foot of a Naturalist has never trod.

With the exception of a few transient Botanists who have worked over, for their own personal pleasure, the neighborhood of some vacation resort the only attempts at obtaining a knowledge of the vegeteble resources of the State may be summarized as follows:

1867 and 1871, Dr. A. S. Todd as Chairman of a Committee of the Medical Society of West Virginia, published a list of the "Medicinal Plants of West Virginia." This list contains an enumeration of 9 trees, 7 shrubs, and 60 herbs.

1870. Mr. DissDebarr, State Commissioner of Immigration in his "Handbook of West Virginia," compiled a list of the timber trees of the State, in which he enumerated 52 species and added 12 species of shrubs.

1876. Prof. Fontaine in compiling his portion of the Centennial volume upon the "Resources of West Virginia," listed more carefully the forest trees, shrubs and medicinal plants of the State, drawing the last from the publication of Dr. Todd. This work contains an enumeration of 69 trees and 16 shrubs.

1878. Profs. H. N. Mertz and G. Guttenberg published a

check list of the "Flora of West Virginia," being an account of work done along the upper Ohio bottoms, and in the mountains of the north-eastern portion of the State, the latter while located at Harper's Ferry. This list enumerate 59 trees, 37 shrubs and 494 herbs.

Miss Verona Mapel, Preceptress of the High School at Glenville, Gilmer county, has quite thoroughly worked over her immediate vicinity in connection with her school duties. She reports 42 trees, 23 shrubs and 290 herbs. Her list does not include the commoner weeds and herbs, nor the grasses or sedges.

1890–92. Mr. L. W. Nuttall an enthusiastic student of structural botany and a keen observer of forms, has spent considerable time working up his locality, (Nuttallburgh, Fayette county). He has kindly furnished me with a manuscript list of about 700 species, many of which I have had the pleasure of examining while visiting at his hospitable home.

Dr. H. McS. Gamble has done considerable herbalizing in connection with his zoological work in Hardy county, near Moorefield. He has kindly donated to the Station his herbarium consisting of 157 species, which are mentioned among the others of this list.

In compiling this Catalogue I have had the assistance afforded by these lists, as well as personal notes from Prof. Guttenberg, Miss Mapel, Mr. Nuttall and Dr. Gamble. I have also been kindly tendered notes taken in the State by Prof. Joseph F. James, Prof. Chas. R. Barnes, Prof. Thos. C. Porter, and Mr. Aug. D. Selby. Prof. Brown, of the University, has also loaned me his herbarium of Glenville plants for examination, as has also Dr. Workman, of Bayard.

All these contributions are acknowledged in the text by a or name initial appened to species or localities. All localities mentioned, where such credit is not given, record my personal observations.

I am under special obligations to Prof. N. L. Britton for his kind help in comparing numbers of my plants with those in the Columbia College Herbarium, as well as for suggestions and assistance in many ways. I am also idebted to Prof. W. A. Kellerman for the examination of nearly all of the epiphytic fungi in this work; and to Prof. Charles H. Peck for the identification of some of the epiphytes, and all of fleshy fungi.

As to nomenclature: I have followed the principle of priority and the double credit system, as far as my access to literature would allow. Omissions in this matter will be found to be in direct proportion to my lack of ability to refer.

My idea of the matter is as follows: Linnæus had not completed his work of generation until the issue of his "Species Plantarum" in 1753. Any changes that he made, therefore, in his previous work, or while classifying the work of others who preceded him, should be acknowledged as positive. Surely a man has a right to correct his own errors. We should not ignore

nore his later ideas by picking out these errors and attempting to establish them as facts today, simply because the errors had the precedence of the correction of them. Linnæus' "Systema Plantarum" (1735) was little more than a mere list of names and should not be accepted as authority in the face of later work done by himself or some other careful systematist, even if he did not change these names in his "Species Plantarum."

To illustrate this, I feel that Robert Brown's Nasturtium should hold; and that Linnæus' Lepidium (1737) cannot be returned to his Nasturtium (1735). Linnæus named one of our genera Pavia in 1735; he corrected this to Esculus in 1737; but in 1753 decided that Æsculus was the proper name for the genus: Is it not right that we should regard this change and acknowledge his correction by using the name Æsculus hereafter? Properly, following this idea, we have no need to fear that the 15,000 species of Astragalus will be changed to Glycia with an "OK" placed after them; how unfortunate it would be to mark with that American symbol for "all correct" many of the changes that Herr Kuntze advises!

As to the double credit system: I judge it no more than right to give the discoverer of any species credit for his work, even if he does err by placing the species in a wrong genus. It is also proper, I think, to acknowledge the student who afterwards detects the error; especially as this acknowledgement is part of the necessary data in considering the species.

Morgantown, W. Va., July, 1892.

RANUNCULACEÆ.

CLEMATIS, L.

C. Virginiana, L.   Virgin's Bower.   M. & G.
River banks, fence rows, etc., Monongalia: Marion:
Preston: Wood. Webster: Long Glade. Gilmer: Glenville
—V. M. Greenbrier: near White Sulphur Springs. Sum-
mers: near Hinton. Kanawha: near Charleston. Fayette:
near Nuttallburg—L. W. N.

C. viorna, L.   Leather Flower.
Thickets in rich soil. Monongalia: Little Falls.
Marion: Valley Falls—K. D. Walker. Fayette: near Nut-
tallburg—L. W. N. Summers: near Hinton.

C. verticillaris, DC.   Mountain Clematis.
Rocky Woods. Monongalia: near Cheat View.

ANEMONE, L.

A cylindrica, Gray.   Long-fruited Anemone.
Dry Woodlands. Wood: near Leachtown. Jackson:
near Sandyville. Rare.

A. Virginiana, L.   Thimble-weed.   M. & G.
Open woods and meadows. Upshur: near Buckhan-
non. Monongalia: along the Monongahela River. Randolph: Rich Mountains, alt. 1,825 ft.. Fayette: near Nuttallburg, where it grows as tall as 4 ft.—L. W. N. Frequent
throughout the State.

A. Canadensis, L.   Pennsylvania Anemone. *A. Pennsylvanica,* L.
Rich woods, rare. Calhoun: along Laurel Run. Gil-
mer: near Glenville—V. M.

A. quinquefolia, L.   Wind-flower. Wood Anemone. *A. ne-
morosa,* L.
Margins of rich woods and opens. Calhoun: along
Laurel Run. Gilmer: near Glenville—V. M. Fayette:

near Nuttallburg—L. W. N. And general throughout the State.

A. trifolia, L.

Rich woods. Mercer: near Ingleside. McDowell: near Elkhorn. Monongalia: near Camp Eden.

The altitude of the first two stations in the southern part of the State is from 2,200 2,350 ft.; these localities are along the same range of mountains as the original station of Canby in Virginia. The Monongalia station in the extreme northern part of the State has an altitude of about 850 ft. and is near Knipes' Pennsylvanian station. Though I have not as yet found the species at any point through the state that would connect these distant points, yet I fully believe that many will be found in the future.

HEPATICA, L.

H. Hepatica (L.), Britt. Hepatica. Liver-leaf. M. & G.

Rocky or rich woods. Gilmer: near Glenville—V. M. Greenbrier: near White Sulphur Springs. Hardy: near Moorefield—G. Mercer: near Bluefield. Monongalia: along Decker's Creek. Fayette: near Nuttallburg, where it often grows in clefts in rocks—L. W. N. And frequent throughout the northern portions of the State.

H. acuta (Pursh). Britt. M. & G.

Rich woods. Gilmer: near Glenville—V. M.—Prof. Brown. Greenbrier: near White Sulphur Springs. Monongalia: near Cheat View. Throughout the State, especially eastward in the mountains. More frequent than the preceding. McDowell: near Elkhorn. Mercer: Bluestone Jc.

SYNDESMON, Hoffm'g.

S. thalictroides (L.), Britt. Rue Anemone. (*Thalictrum anemonoides*, Michx.) M. & G.

Woods and hillsides. Gilmer: near Glenville—V. M. Fayette: near Nuttallburg—L. W. N. Monongalia: near Uffington and Morgantown. Frequent throughout the State. Hardy: near Moorefield—G. Mercer: near Beaver Spr.

Forma rosea.

A beautiful rose colored clump of the species has persisted for two years in Monongalia at Little Falls near the Cascade.

THALICTRUM, L.

T. dioicum, L. · M. & G., V. M., L. W. N.
Rocky woods: frequent throughout the State. Especially noticeable in the Alleghanies.

T. polygamum, Muhl. Common Meadow Rue. M. & G.
Damp meadows and near streams and ditches. Randolph: rich growths in the bottoms along Tygart's Valley River. Fayette: near Kanawha Falls—James; near Nuttallburg, alt. 2,000 ft.—L. W. N. Gilmer: near Glenville. Frequent throughout the State.

T. purpurascens, L. Purplish Meadow Rue.
Dry, open woods and rocky hillsides. Wirt: near Elizabeth; near Burning Springs. Webster: near Upper Glade. Randolph: along Tygart's Valley River, alt. 2,100 ft. Fayette: near Nuttallburg – L. W. N.

T clavatum, DC.
Fayette: Near Nuttallburg, in sandy clefts of rocks around waterfall, alt. 200 ft., one station—L. W. N.

TRAUVETTERIA, F. & M.

T. palmata, F. & M. False Bug-bane.
Plentiful along the Blackwater Fork of Cheat, about one mile below Davis in Tucker county. Fayette: near Hawk's Nest; and Loup Creek—James; near Nuttallburg, common—L. W. N.

RANUNCULUS, L.

R. ambigens, Wats. Water Plantain Spear Wort (R. *alismaefolius*, Gray.)
One station only; Upshur: in a marshy spot near Lorentz.

R. abortivus, L. Small-flowered Crowfoot. L. W. N., V. M., M. & G.
Damp, shady places, frequent throughout the State.

R. sceleratus, L. Cursed Crowfoot. V. M.
Moist places, common, throughout the State.

R. recurvatus, Poir. Hooked Crowfoot. M. & G.
Rich, open woods. Abundant along the Monongahela river in Upshur, Barbour, Taylor, Marion, and Monongalia counties. Fayette: near Nuttallburg, not common—L. W. N. Mercer: near Simmons.

R. fascicularis, Muhl. Early Crowfoot.
>Dry or moist grassy hillsides. Monongalia: near Morgantown. Mineral: near Keyser—W. Gilmer: near Glenville –V. M. Hardy: near Moorefield—G.

R. septentrionalis, Poir. Buttercup.
>Moist, shady places. Fayette: near Nuttallburg – L. W. N., and general throughout the State.

R Pennsylvanicus, L. Pennsylvania Buttercup.
>Damp woods. Monongalia and Marion, along the Monongahela river.

R. REPENS, L. Creeping Crowfoot. M. & G.
>Low grounds. Mineral: Banks of the Potomac near Keyser. Jefferson: near Shenandale Springs.

R. ACRIS, L. Tall Buttercups.
>Waste places infrequent. Wood: near Parkersburg, Jefferson: near Shenandoah Junction. Randolph: in clearings on Cheat Mountain, near Cheat Bridge, alt. 3350 ft. Gilmer: near Glenville—V. M.

HELLEBORUS, L.

H. VIRIDIS, L. Green Hellebore.
>Hardy: near Moorefield —G. It was from specimens sent to Dr. Gray from this station by Dr. Gamble, that the location "W. Va." was credited in the Manual.

CALTHA, L.

C. palustris, L. Marsh Marigold.
>Two stations only known to me; Grant: in a spring run in deep woods near Bayard, about fifty plants within an area of one hundred feet. Preston: near Terra Alta.

ISOPYRUM, L.

I. trifolium, (L.), Britt. Gold Thread. *Coptis trifolia*, Salisb.
>Deep, rich mountain woods. Preston: near Mill Run and Cranesville.

AQUILEGIA, L.

A. Canadensis, L. Wild Columbine. M. & G.
>Damp, rocky places. Mineral: along Knobby mountains. Monongalia: along Cheat River. Calhoun: along Little Kanawha River. Gilmer: near Glenville—V. M.;

Prof. Brown. Fayette: near Nuttallburg—L. W. N. Hampshire: near Does Gully. Hardy: near Moorefield. A small form 6 to 8 inches high, with small leaves and flowers. Mercer: near Beaver Spr.

DELPHINIUM, L.

D. tricorne, Michx. Dwarf Larkspur.
Dry woods. Monongalia: along Cheat and Monongahela Rivers. Marion: along the Monongahela. Gilmer: near Glenville—V. M.; Prof. Brown. Frequent throughout the northern part of the State.

Forma **albiflora.**
Monongalia: prevalent near Stumptown. The flowers are pure white with no tinge of blue.

D. CONSOLIDA, L. Field Larkspur.
Frequent in old fields and along roadsides. Lewis: along Stone Coal Creek. Monongalia: near Morgantown. Upshur: near Buckhannon. Randolph: near Beverly. Kanawha: near Kanawha Falls—James.

ACONITUM, Tourn.

A. uncinatum, L. Wild Monk's Hood.
Deep, rich woods along streams. Randolph: along Cheat River near Cheat Bridge. Monongalia: near Camp Eden. Fayette: near Nuttallburg, along New River—L. W. N.

CIMICIFUGA, L.

C. Americana, Michx. American Bug-bane.
Plentiful throughout the mountain regions of Mineral, Hampshire, Grant, Tucker. Hardy: near Moorefield; Randolph, Pendleton, Webster, Pocahontas, and Greenbrier counties. Fayette: near Nuttallburg—L. W. N. McDowell: near Elkhorn.

C racemosa (L.), Nutt. Rattle-weed, Black Cohosh. M. & G.
Rich opens and clearings. Wood: near Leachtown. Lewis: along Stone Coal Creek. Randolph: near Valley Bend; Point Mountain, alt. 3300 ft. Webster: Buffalo Bull Mountains. Gilmer: near Glenville—V. M. Fayette: near Nuttallburg—L. W. N. and frequent throughout the State.

ACTÆA. L.

A. spicata, (L.) var. **rubra,** Ait. Red Baneberry.
Rich woods, infrequent. Randolph: on Point Mountain. Grant: near Bayard. Tucker: near Davis.

A. alba. Mill. White Baneberry. M. & G.
Rich woods. Randolph: on Point Mountain. Tucker:
near Davis. Grant: near Bayard. Fayette: near Nuttallburg.—L. W. N.

HYDRASTIS, L.

H. Canadensis, L. Yellow Root, Yellow Puccoon, Golden Seal.
Deep Rich woods. Jackson: near Ripley. Wirt near
Burning Springs. Grant: near Bayard. Upshur: summit
on Staunton pike. Lewis: along Leading Creek. Calhoun:
along Laurel Run: Gilmer: near Glenville.—V. M.; Prof.
Brown. Monongalia: along Cheat River.

XANTHORRHIZA, Marshall

Z. apiifolia. L. Her. Shrub Yellow Root.
Rich rocky mountain woods. Nicholas: on Mumble-
the-Peg Creek, plentiful. Fayette: near Nuttallburg, common.—L. W. N.

MAGNOLIACEÆ.

MAGNOLIA, L.

M. acuminata, L. Cucumber Tree, "Yellow Linn." M. & G.
Rich woods. Monongalia: along the Monongahela
river, especially near Little Falls, Opekiska, and Montana.
Wirt: along Little Kanawha River. Randolph: on Point
Mountain, alt. 2335–3700 ft. Webster: along Buffalo Bull
Mountains. Nicholas: at Beaver Mills, and Collett's Glade.
Jackson: near Sandyville. Mineral: near Keyser. Preston:
along B. & O. R. R. Randolph: on Cheat Mountain, alt.
2800 ft. Gilmer: near Glenville—V. M. Monroe: near Alder-
son. Summers: near Greenbrier Stock Yards; near Hinton.
Kanawha: near Handley. Harrison: near Shinnston and
Lumberport. Fayette: near Nuttallburg—L. W. N.

M. tripetala, L. Umbrella tree. *M. Umbrella,* Lam.
Rich woods near streams. Wirt: near Burning
Springs. Randolph: on Point Mountain, alt. 2800 ft. Ka-
nawha: near Charleston—C. R. Barnes; James. Nicholas:
near Beaver Mills, alt. 2125 ft. Gilmer: near Glenville—V.
M. Monongalia: Little Falls. Summers: near Hinton.
Harrison: near Shinnston and Lumberport. Fayette:
near Nuttallburg, rare—L. W. N. Hardy: near Moorefield.
McDowell: near Elkhorn. Mercer: Bluestone Junction and
Ingleside.

M. Fraseri, Walt. Ear-leaved Magnolia.
Deep rich mountain woods. Randolph: on Point
Mountain, alt. 3700 ft. Webster: on Buffalo Bull Mountain,
alt. 3400 ft. Nicholas: near Beaver Mills, alt. 2125 ft. Fay-
ette: near Nuttallburg – L. W. N. Summers: near Hinton.
McDowell: near Elkhorn. Mercer: Bluestone Jc.

LIRIODENDRON, L.

L. Tulipifera L. Tulip Tree, "White, Yellow, or Hickory Pop-
lar." M. & G.
Common in rich woods throughout the State. Mag-
nificent trees in Randolph, Pocahontas, Greenbrier, Webster,
Nicholas. and Preston counties. Summers: near Hinton.
Fayette: near Kanawha Falls; near Nuttallburg, common—
L. W. N. Hardy: near Moorefield—G. Mercer: near
Ingleside.

ANONACEÆ.

ASIMINA, Adans.

A. triloba (L.), Dunal. Papaw. M. & G.
Rich soil near streams, common. Wood: near Kanawha Station. Wirt: along Straight Creek. Webster: Buffalo Bull Mountains, alt. 2100 ft. Nicholas: along Peter
Creek. Fayette: along Gauley river. Kanawha, Putnam
and Mason: along the Great Kanawha. Gilmer: near Glenville—V. M. Greenbrier: near Ronceverte. Summers: near
Hinton. Marion: near Worthington, Fairmont and Montana.
Monongalia: general. Fayette: near Nuttallburg, common.
– L. W. N. Morgan: near Cacapon. Hardy: near Moorefield. Mercer: near Ingleside.

MENISPERMACEÆ.

MENISPERMUM, L.

M. Canadensis, L. Moonseed. V. M., M. & G.
Thickets. Wirt: along Little Kanawha River. Marion and Monongalia: along the Monongahela. Greenbrier:
near White Sulphur Springs. Fayette: near Nuttallburg,
pistillate plants, rare.—L. W. N.: and frequent along streams
throughout the State. Hardy: near Moorefield. Mercer:
near Princeton.

BERBERIDACEÆ.

BERBERIS, L.

B. Canadensis, Pursh. Barberry.
Fields and roadsides. Mercer: near Beaver Spring,
where it partakes of the nature of a weed.

B. VULGARIS, L.
Thickets and roadsides. Monongalia: a wide escape
near Laurel Point. Mercer: a wide escape near Ingleside.

CAULOPHYLLUM, Michx.

C. thalictroides, (L.), Michx. Blue Cohosh. M. & G.
Deep, rich woods. Randolph: Rich Mountains; Point
Mountain, alt. 3,300 ft. Monongalia: along the Mononga-
hela River. Gilmer: near Glenville—V. M. Fayette: near
Nuttallburg—L. W. N. Mercer: near Bluefield. Frequent
throughout the State.

JEFFERSONIA, Barton.

J. diphylla, (L.), Pers. Twin-leaf. M. & G.
Rich woods. Monongalia: near Morgantown. Marion: near Glover's Gap. Wetzell: near Littleton. Marshall:
near Glen Easton. Gilmer: near Glenville—V. M.; Prof.
Brown. Cabell: near Huntington—Aug. Selby.

PODOPHYLLUM, L.

P. peltatum, L. May Apple. Mandrake. V. M.; M. & G.
Frequent throughout the State, in some rich spots
very abundant. Randolph: Rich Mountains, alt. 1610–
2125 ft.; Point Mountain, alt. 3300 ft. Fayette: near Nuttallburg—L. W. N. Hardy: near Moorefield—G. Mercer:
Bluestone Jc.

NYMPHACEÆ.

CASTALIA, Salisb.

C. odorata, (Dryand). Woodv. & Wood. White Water Lily.
Nymphaea odorata, Ait.
In slack waters. Preston: near Pennsylvania line,
rare.

NYMPHÆA, L.

N. advena, Soland.   Yellow Pond Lily.   V. M.; M. & G.
In slack waters. Preston: near Terra Alta. Wood:
Wirt: Calhoun and Gilmer: along the Little Kanawha
River. Upshur: near Lorentz. Morgan: along the Little
Cacapon. Putnam; near Buffalo. Hardy: near Moorefield.

PAPAVERACEÆ.

SANGUINARIA, L.

S. Canadensis, L.   Blood-root. "Coon-root."   M. & G.
Rich woods, frequent. Randolph: Point Mountain.
Monongalia: near Morgantown. Jefferson: near Flowing
Spring, and Shenandale Spring. Gilmer: near Glenville—
V. M. Fayette: near Nuttallburg—L. W. N. Throughout
the State. Hardy: near Moorefield—G.

STYLOPHORUM, Nutt.

S. DYPHLLUM (Michx.) Nutt.   Celandine Poppy.
Old fields. Ohio: near Wheeling—M. & G.

CHILIDONIUM, L.

C. MAJUS, L.   Celandine.   M. & G.
Waste grounds; near dwellings. Monongalia: near
Easton; near Morgantown. Jefferson: near Charlestown,
abundant. Lewis: near Weston. Hardy: near Moorefield.—
G. Hampshire: near Romney.

PAPAVER, L.

P. DUBIUM, L.   Smooth-fruited Corn-popy.
Cultivated grounds, and waste fields. Jefferson: near
Shenandoah Je.; near Charlestown, abundant. Berkeley:
near Hedgesville, a weed.

FUMARIACEÆ.

ADLUMIA, Raf.

A. fungosa, Ait. Greene.
Monongalia: climbing over rocks, Tibb's Run.

BICUCULLA, Adans.

B. Cucullaria, (L.) Dutchman's Breeches. M. & G.
Rich woods. Monongalia: near Morgantown; along Cheat River. Gilmer: near Glenville—V. M. Hardy: near Moorefield—G. Fayette: near Nuttallburg—L. W. N.

B. Canadensis, (Goldie.) Squirrel Corn. M. & G.
Rich woods. Monongalia and Marion: along the Monongahela River. Preston: along Cheat River. Fayette: near Nuttallburg—L. W. N.

B. eximina, (DC.)
Pocahontas: summit of Spruce Knob, alt. 4800 ft.—A. D. Hopkins.

NECKERIA, Scop.

N. glauca, (L.) Pale Corydalis, M. & G.
On rocks. Randolph: on Lone Sugar Knob. Gilmer: near Glenville—V. M.

N. flavula, (Raf.) Yellow Corydalis.
Rich soils. Common throughout the northern counties. Fayette: near Nuttallburg—L. W. N. Hardy: near Moorefield.

N. aurea, (Willd.) Golden Corydalis.
Along streams. Hardy: near Moorefield—G.

FUMARIA, L.

F. OFFICINALIS, L. Fumitory.
Waste places. Hardy: near Moorefield—G.

CRUCIFERÆ.

NASTURTIUM, R. Br.

N. OFFICINALE, R. Br. Water Cress. M. & G.
Cold spring runs, frequent. Jefferson: near Shenandale Springs; near Flowing Spring. Lewis: on Leading Creek. Wirt: near Elizabeth.

N. SYLVESTRE, (L.), R. Br. Yellow Wood-cress.
Moist places in open woods, frequent. Monongalia: near Morgantown. Preston: Cold Spring. Marion: near Montana. Jefferson: near Shenandale Springs.

N. obtusum, Nutt.
River banks. Mason: near Point Pleasant.

N. palustre, (L.), DC. Marsh Cress. M. & G.
Marshy places and glades, infrequent. Webster: near Welch Glade. Kanawha: near Charleston. Preston: near Kingwood. Fayette: near Nuttallburg, rare—L. W. N. Mason: near Point Pleasant; Banks of the Ohio. Wood: near Parkersburg.

N. hispidum, (Desv.), DC. *N. palustre* var. *hispidum*, Gray. Preston: near Kingwood.

N. ARMORACIA, (L.), Fries. Horseradish. M. & G.
Escaped from cultivation in many waste places and *fields. Marshall: frequent in several places where it is complained of as a weed; difficult to eradicate. Jefferson: near Shenandoah Junction.

BARBAREA, R. Br.

B. VULGARIS, R. Br. Yellow Rocket. M. & G.
Becoming a weed in many places in Jefferson: Berkeley: Morgan: Hardy: near Moorefield—G; and other counties.

B. PRÆCOX, (Smith), R. Br, Scurvy Grass.
Running wild near Charlestown in Jefferson; and Lewis: near Weston.

ARABIS. L.

A. patens, Sulliv.
Moist rocky places in woods. Monongalia: near Monongahela River at Uffington. Preston: Cold Spring.

A. lævigata, (Muhl.), Poir. M. & G.
Rocky places, frequent. Monongalia: near Morgantown, and Little Falls. Marion: near Catawba. Fayette: near Nuttallburg, common—L. W. N.

A. Canadensis, L. Sickle-pod. M. & G.
Woods, and along cool runs. Monongalia: near Granville, near Morgantown and Uffington.

A. lyrata, L. M. & G.
Rocky or sandy places. Monongalia: shores near mouth of Cheat River. Mercer: near Ingleside.

CARDAMINE, L.

C diphylla, (Michx.), Wood. Pepper Root. *Dentaria diphylla*, Michx. M. & G.
Common in deep, cool ravines and in the mountains. Monongalia: Wirt: Wood: Calhoun: Lewis: Upshur: Jef-

ferson: Grant: and Tucker counties. Gilmer: near Glenville—V. M.; Prof. Brown. Fayette: Hawk's Nest—Porter; near Nuttallburg—L. W. N.

C. heterophylla, (Nutt.), Wood. *D. heterophylla*, Nutt.
Rocky, moist places. Monongalia, near Little Falls; opposite Uffington. Fayette: near Nuttallburg—L. W. N.

C. laciniata, (Muhl.), Wood. M. & G.
Moist woods, frequent. Monongalia: opposite Beechwoods and Uffington; near Little Falls. Preston; Cold Spring and elsewhere. Gilmer: near Glenville—V. M. Fayette: near Nuttallburg—L. W. N. Hardy: near Moorefield—G.

Var. multifida, (Muhl.)
Rich woods. Monongalia: near Little Falls.

C. hirsuta, L. Small Bitter Cress. M. & G.
Wet places. Monongalia: Monongahela River below Morgantown; above Little Falls. Marion: near Catawba and elsewhere. Fayette: near Nuttallburg—L. W. N.

C. bulbosa, (Schreb). B. S. P. Spring Cress. *C. rhomboidea*, DC. M. & G.
Wet meadows and springy places. Preston: Cold Spring. Monongalia: road to Cheat River beyond Easton. Wood: Kanawha Station. Fayette: near Nuttallburg, common—L. W. N. Mercer: near Bluefield.

C. Douglassii, (Torr.), Britt. (*C. rhomboidea*, var. *purpurea*, Torr.)
Damp places. Monongalia: near Morgantown.

C. rotundifolia, Michx. Mountain Water Cress. M. & G.
Cool Springs. Preston: Cold Spring. Jefferson: Flowing Spring. Wirt: near Burning Springs. Calhoun: Laurel Run. Monongalia; near Morgantown.

DRABA, L.

D. ramosissima, Desv. Whitlow Grass.
On wet cliffs. Jefferson: cliffs along Shenandoah River between Millville and Harper's Ferry. Tucker: Cliffs near Falls of Blackwater. Hardy: near Moorefield.

D. verna, L. Shad Flower.
Sandy wastes and roadsides. Monongalia: banks of Falling Run; banks Monongahela below Morgantown, and near Little Falls. Marion: near Opekiska. Fayette: near Nuttallburg—L. W. N,

HESPERIS. L.

H. MATRONALIS. L. Dame's violet.
Escaped to waste places. Monongalia: cinders of railroad banks near Morgantown.

SISYMBRIUM, L.

S. OFFICINALE, (L.), Scop. Hedge Mustard. M. & G.
Roadsides and ditches, too common throughout the State.

S. THALINA, (L.), Gay.
Waste grounds. Fayette: near Nuttallburg—L. W. N.

ERYSIMUM, L.

E. CHEIRANTHOIDES, L. Worm-seed Mustard. M. & G.
Roadsides and railroad embankments. Monongalia: near Morgantown. Jefferson: near Shenandoah Junction. Mineral: near Piedmont.

CAMELINA, Crantz.

C. SATIVA, (L.), Crantz. False Flax.
Fields and waste grounds. Mineral: near Keyser— W. Jefferson: near Charlestown; near Shenandoah Jc.

BRASSICA, L.

B. NIGRA, (L.), Koch. Black Mustard. M. & G.
Fields and waste places. Fayette: near Nuttallburg, much eaten as "greens" in spring—L. W. N. A common weed throughout the State.

B. SINAPISTRUM, Boiss. Charlock. "Crowd-weed." Kraut-weed.
A miserable weed in wheat fields in Jefferson and Berkeley counties. Less abundant elsewhere throughout the State.

CAPSELLA, Moench.

C. BURSA-PASTORIS, (L.), Moench. Shepherd's Purse. L.W.N., M. & G.
Fields and roadsides, common throughout the State.

LEPIDIUM, L.

L. Virginicum, L. Wild Peppergrass. L. W. N., M. & G.
Fields and roadsides, common throughout the State.

L CAMPSTRE (L.), R. Br.   English Peppergrass, "Glenn-weed," "Glenn-pepper," "Crowd-weed."
An exceedingly abundant weed in Jefferson and Berkeley counties, where it is known as "Glenn-weed," it being first noticed in the fields of Colonel Glenn; who tells me that the weed was quite plentiful, however, in these fields before he purchased them, having been brought there in clover seed bought in Hagarstown, Md., and sown by the previous owner of the farm. The weed is now the worst pest in the large wheat fields of those counties.

RAPHANUS, L.

R. SATIVUS, L.   Radish.
Frequently persistent in waste grounds and cultivated fields, in many parts of the State.

CAPPARIDEÆ.

CLEOME, L.

C. SPINOSA, L.   Spider Flower.   *C. pungens* Willd.
Escaped from farther south, at Barboursville near the Guyandootte River, Cabell county—Prof. James, 1877.

CISTINEÆ.

HELIANTHEMUM, Pers.

H majus, (L.) B. S. P.   Frost-weed.   *H. Canadense,* Michx.
Dry soils.   Preston: near Terra Alta.

LECHEA, L.

L. minor, L.   Pin-weed.   *L. major.* Michx.
Dry places.   Summers: near Hinton.   Fayette: near Nuttallburg, alt. 1600 ft.—L. W. N.

L Leggettii, Britt & Hollick.   *L. minor,* Lam.
Dry sandy places.   Fayette: near Nuttallburg, plentiful at an alt. of 2000 ft.—L. W. N.

VIOLARIEÆ.

VIOLA, L.

V. pedata, L.   Bird's-foot Violet.
Sandy soils.   Monongalia: at The Flats.   Mineral: along the Potomac near Keyser—W.   Randolph: on Point Mountain.

Var. **bicolor**, Pursh. Pansy Violet.
Hardy: near Moorefield—G. Hampshire: near Romney —A. D. Hopkins.

V. palmata, L. M. & G.
Near runs in moist ground. Monongalia: near The Flats, Morgantown. Gilmer: near Glenville—V. M. Fayette: near Nuttallburg—L. W. N.

V. cucullata, Ait. Common Blue Violet. L. W. N., V. M., M. & G., G.
Low grounds; common throughout the State.

V. sagittata, Ait. Arrow Leaved Violet. M. & G.
Dry or moist, sandy places. Monongalia: near Morgantown, Uffington and Little Falls. Preston: near Cold Spring. Gilmer: near Glenville—V. M. Fayette: near Nuttallburg—L. W. N.

V. blanda Willd. Sweet White Violet. M. & G.
Damp woods. Monongalia; near Morgantown. Gilmer: near Glenville—V. M. Fayette: near Nuttallburg—L. W. N., and frequent throughout the State. McDowell: near Elkhorn.

V. primulaefolia, L.
Damp soils. Fayette: near Nuttallburg—L. W. N. McDowell: near Elkhorn.

V. lanceolata, L. Lance-leaved Violet.
Boggy places. Monongalia: up Falling Run, above Morgantown, the only station so far known to me.

V. rotundifolia, Michx. Round-leaved Violet. M. & G.
Cold woods. Randolph: Rich Mountains, alt. 2110 ft. Gilmer: near Glenville—V. M. Fayette: near Nuttallburg—L. W. N. Mercer: near Ingleside and Wills.

V. pubescens, Ait. Yellow Violet. M. & G.
Rich woods. Mineral: near Keyser—W. Randolph: Rich Mountains, alt. 2125 ft. Grant: near Bayard. Tucker: along Blackwater Fork of Cheat. Gilmer: near Glenville—V. M. Fayette: near Nuttallburg—L. W. N. Monongalia: near Cassville.

Var. **scabriuscula**, T. & G.
Rich woods. Monongalia: near Morgantown, common. Fayette: near Nuttallburg—L. W. N.

V. hastata, Michx.
> Woodlands. Fayette: near Nuttallburg—L. W. N.
> Oak woods. Mercer: near Bluefield. McDowell: near Elk-
> horn.

V. Canadensis. L. Canada Violet. M. & G.
> Rich woods. Monongalia: magnificent specimens in
> great profusion along the woods bordering the F. M. &
> P. R. R., between Beechwoods and Little Falls; near Ulling-
> ton. Marion: near Opekiska and Catawba. Gilmer: near
> Glenville V. M. Fayette: near Nuttallburg - L. W. N.
> McDowell: near Elkhorn.

V. striata, Ait. Pale Violet. M. & G.
> Along runs. Monongalia: the most common species.
> Fayette: near Nuttallburg—L. W. N. Hardy: near Moore-
> field—G. McDowell: near Elkhorn.

V. rostrata, Muhl. Long-spurred Violet.
> Hillside. Fayette: near Nuttallburgh—L. W. N.

V. tenella. Muhl. Field Violet. (*V. tricolor, var. arvensis*, DC.)
> Fields and rocky opens. Monongalia:common. Min-
> eral: near Keyser—W. Fayette: near Nuttallburg—L.
> W. N.

SOLEA, Ging.

S. concolor (Forst.), Ging. Green Violet. M. & G.
> Rich woods. Wirt: near Burning Springs. Calhoun:
> along Laurel Run. Gilmer: near Glenville—V. M.

POYLGALEÆ.

POLYGALA, L.

P. sanguinea, L. Red Milkwort.
> Sandy fields. Wood: near Kanawha Station. Web-
> ster: Upper Glade. Preston near Terra Alta. Fayette: near
> Nuttallburg, alt. 2400 ft.—L. W. N.

Forma. **albiflora**.
> In the glades of Webster and Preston counties.

P. mariana. Mill. *P. fastigiata, Nutt.*
> Damp places. Preston: near Terra Alta.

P. Curtissii, Gray.
> Glady spots. Fayette: near Nuttallburg—L. W. N.

P. cruciata, L.   Cross Milkwort.

Margins of Glades.   Preston: near Reedsville.   Webster: Upper Glade.

P. verticillata, L.   Whorled Milkwort.

Dry places.   Wirt: near Burning Springs.   Lewis: near Leading Creek.   Upshur: near Buckhannon.   Summers: near Hinton.

P. ambigua, Nutt.

Dry soils.   Wood: near Lockhart's Run.   Wirt: near Elizabeth.   Calhoun: on Nigh-Cut Hill.   Monongalia: hills below Morgantown.   Fayette: near Nuttallburgh—L. W. N.

P Nuttallii, T. & G.

Mountain woods.   Fayette: near Nuttallburgh, alt. 2000 ft.—L. W. N.

P. Senega, L.   Seneca Snake-root.

Rocky soils and rich bottoms.   Mineral: near Keyser—W.   Webster: in Welsh, Long and Collett's glades.   Preston: Reedsville glade.   (Long glade is said to be white with this species when in bloom.)

P. paucifolia, Willd.   Fringed Polygala.

Rocky woods.   Mineral: near Keyser along the Knobby Mountains.—W.   Hardy: near Moorefield—G.

CARYOPHYLLEÆ.

DIANTHUS, L.

D. ARMERIA, L.   Deptford Pink.

Fields, roadsides, and river banks.   Marion: near Catawba—K. D. Walker.   Gilmer: near Glenville—V. M.   Fayette: near Nuttallburg, rare—L. W. N.; near Kanawha Falls—James.   Summers: near Hinton.   Jefferson: at Harper's Ferry—M. & G.

SAPONARIA, L.

S. OFFICINALIS, L.   Soap-wort.   Bouncing Bet.

Becoming a very common weed along roadsides throughout the more settled portions of the State.   In especially large areas along the B. & O. R. R. and Shenandoah Valley R. R., in Jefferson Co.   Calhoun: Grantsville.   Gilmer: Glenville—V. M.   Berkeley: Martinsburg.   Summers: near Hinton.   Fayette: near Nuttallburg—L. W. N.   Hardy: near Moorefield—G.   Monongalia: near Lock 9.

SILENE,L .

S. stellata, (L.), Ait.   Starry Campion.   M. & G.
Wooded banks, frequent. Wood: Wirt: Calhoun:
Gilmer: Lewis: and Upshur, common. Preston: near Ter-
ra Alta. Summers: near Hinton. Greenbrier: near White
Sulphur Springs. Fayette: near Nuttallburg—L. W. N.
Hardy: near Moorefield—G.

S. Virginica, L.   Fire Pink. Catchfly.   M. & G.
Open woods. Kanawha: near Charleston—Barnes.
Fayette: near Nuttallburg, very common—L. W. N. Mer-
cer: near Bluefield. Monongalia: permian formations at
Cassville.

S. nivea, Otth.
Wooded banks.   Fayette: near Nuttallburg—L.W.N.

S. Pennsylvanica, Michx.   Wild Pink.
Open woods. Monongalia: near Morgantown. Min-
eral: near Keyser—W. Gilmer: near Glenville—V. M.
Hampshire: Doe's Gully.

S. antirrhina, L.   Sleepy Catchfly.
Dry places.   Fayette: near Nuttallburg—L. W. N.

LYCHNIS, L.

L GITHAGO, (L.), Lam.   Corn Cockle.   V. M.: M. & G.
A frequent weed in wheat fields throughout the State.
Fayette: near Nuttallburg, in shady places—L. W. N.

CERASTIUM, L.

C. VULGATUM, L.   Large Mouse-ear Chickweed.
Waste grounds and fields. Ohio: Cowan's Hill, near
Wheeling—M.& G. Gilmer: near Glenville—V. M. Wood:
near Waverly and elsewhere, becoming frequent. Fayette:
near Nuttallburg—L. W. N.

C. VISCOSUM, L.   Mouse-ear Chickweed.
Fields. Ohio: Cowan's Hill, near Wheeling—M. &
G. Fayette: near Nuttallburg—L. W. N. Monongalia:
near Morgantown.

C. arvense, L.   Field Chickweed.   M. & G.
Dry places. Frequent throughout the State.

C nutans, Raf.   Star Chickweed.   M. & G.
Woods near streamlets; common in Monongalia and
Marion counties.

STELLARIA, L.

S. MEDIA, (L.), Smith. Chickweed. V. M., M. & G.
Damp places, common everywhere. Fayette: near
Nuttallburg, blooms all winter—L. W. N.

S. pubera, Michx. Great Chickweed. M. & G.
Shady places, common. Monongalia, Marion and
Preston counties. Gilmer: near Glenville—V. M. Fayette:
near Nuttallburg—L. W. N. Mercer: near Bluefield.

S. longifolia, Muhl. Long-leaved Stitchwort.
Damp soils. Gilmer: near Glenville— V. M.

ARENARIA, L.

A. SERPYLLIFOLIA, L. Thyme-leaved Sandwort. M. & G.
Sandy banks. Kanawha: near Charleston—Barnes.
Hardy: near Moorefield—G.

SPERGULA, L.

S. ARVENSIS, L. Field Spurry.
Fields. Preston: near Cranberry Summit—M. & G.;
near Terra Alta.

PORTULACEÆ.

PORTULACA, L.

P. OLERACEA, L. Purslane. "Pussley." L. W. N., M. & G.
A weed in cultivated grounds and gardens. Common
throughout the State.

CLAYTONIA, L.

C. Virginica, L. Spring Beauty. M. & G.
Common throughout the northern parts of the State,
in rich open woods and along spring runs. Gilmer: near
Glenville—V. M.; Prof. Brown. Hardy: near Moorefield
—G.

C. Caroliniana, Michx.
Frequent with the former species. Marion, Preston,
Wood, Wirt, Calhoun, Gilmer, Lewis, Upshur and Randolph
counties. Fayette: near Nuttallburg—L. W. N.

HYPERICINEÆ.

ASCYRUM, L.

A. Crux-Andreae, L.   St. Andrew's Cross.   M. & G.
  Dry sandy places. Upshur: summit of Staunton Pike.
Randolph: Rich Mountains; along Tygart's Valley River.
Fayette: near Gauley Bridge; near Nuttallburg—L. W. N.
Cabell: near Barboursville—James.

HYPERICUM, L.

H. prolificum, L.   M. & G.
  Glade regions. Webster: Upper Glade. Preston: near
Reedsville. Gilmer: near Glenville—V. M. Fayette: near
Kanawha Falls—James; near Nuttallburg—L. W. N.

H. densiflorum, Pursh.   Shrubby St. John's Wort.   M. & G.
  Glade regions,and moist meadows. Wood: near Lockhart's Run. Webster: in the glades.' Preston: Terra Alta,
and Reedsville glades. Fayette: near Nuttallburg—L.W. N.

H. virgatum, Lam. var. **actuifolium,** Coulter.
  Fayette: near Hawk's Nest—Porter; near Nuttallburg—L. W. N.

H. PERFORATUM, L.   St. John's Wort.   "St. John."   M. & G.
  Fields and roadsides. Randolph: along Tygart's Valley River. Monongalia, Marion, Preston; Grant: near Bayard. Fayette: near Nuttallburg—L. W. N. Not yet very
plentiful in the State.

H. maculatum, Walt.   Spotted St. John's Wort.   M. & G.
  Glade regions,and wet places. Wood: near Kanawha
Station. Wirt: near Elizabeth. Preston: near Reedsville
and Terra Alta. Webster: Upper, Long, and Welsh Glades.
Fayette: near Hawk's Nest and Kanawha Falls—James;
near Nuttallburg—L. W. N.

M. mutilum, L.   L. W. N., M. & G.
  Ditches and low grounds, common throughout the
State.

H. Canadense, L.   Canadian St. John's Wort.
  Glade regions of Preston and Webster counties. Fayette: near Nuttallburg, in spahgnum bogs—L. W. N.

H. gentianoides, (L.), B. S. P.   Orange Grass.   (*H. Sarthora,*
                                      Michx.
  Dry fields.   Monongalia: near The Flats.   Wood:

near Kanawha Station. Fayette: near Kanawha Falls—James.

H adpressum, Barton.
Moist grounds. Greenbrier: near White Sulphur Springs.

H ellipticum, Hook.
In sphagnous glades. Preston: near Terra Alta.

MALVACEÆ.

ALTHÆA, L.

A. ROSEA, L. Hollyhock.
Appears annually along the B. & O. R. R. tracks in Berkeley: near North Mountain, apparently a thorough establishment.

MALVA, L.

M. ROTUNDIFOLIA, L. Common Mallow. G., L. W. N., M. & G.
Cultivated grounds and waysides, a common weed.

M. MOSCHATA, L. Musk Mallow.
Roadsides and meadows escaped from cultivation. Lewis: along Stone Coal Creek. Upshur: near Lorentz. Monongalia: near Morgantown.

SIDA, L.

S. SPINOSA, L.   M. & G.
Waste grounds and fields. Monongalia: near Morgantown, common. Fayette: near Nuttallburg—L. W. N. Jefferson: near Shepherdstown. Mason: near Point Pleasant.

S. Napæa, Cav.
Rocky banks along the Great Kanawha River. Kanawha: opposite Cannelton. Fayette: Quinnimont; near Nuttallburg, frequent and always with 8 carpels—L. W. N. Mason: near Point Pleasant.

ABUTILON, Gaertn.

A. AVICENNE, Gaertn. Indian Mallow. American Jute.
A too common weed in waste and cultivated soils. Monongalia, Wood, Wirt, Calhoun; Mason: near Point Pleasant; near Brighton. Fayette: near Nuttallburg—L. W. N. Berkeley: near Martinsburg, a very bad weed.

HIBISCUS, L.

H. Moscheutos, L. Swamp Rose Mallow.
Brackish, marshy places, and ditches near salt works.
Mason: near Point Pleasant. Kanawha: near Charleston.
Fayette: near Nuttallburg—L. W. N. The pink form,
Jackson: near Sandyville. Hardy: near Moorefield.

H. Trionum, L. Bladder Ketmia. M. & G.
Cultivated grounds. Monongalia: a weed in our experimental plats.

TILIACEÆ.

TILIA, L.

T. Americana, L. Linden. Basswood. V. M., M. & G.
Rich woods. Gilmer: at DeKalb P. O. Randolph:
on Point Mountain. Grant: near Bayard. Monongalia:
near Morgantown, Uffington and Little Falls. Mason: near
Point Pleasant. McDowell: near Elkhorn.

T. heterophylla, Vent. White Basswood. "Lin."
Deep woods. Jefferson: near Charlestown, near Flowing Spring Mill. Fayette: near Nuttallburg—L. W. N.

LINEÆ.

LINUM, L.

L. Virginianum, L. Wild Flax.
Open woods, borders and roadsides. Wood, Wirt,
Calhoun, Gilmer, Lewis and Upshur. Randolph: along Tygart's Valley River. Webster and Nicholas counties. Kanawha: near Peabody; near Coalburgh—James. Jackson,
Monongalia, and Preston. Fayette: near Gauley Bridge;
near Kanawha Falls—James; near Nuttallburgh—L. W. N.
Kanawha: near Charleston—James.

L. striatum, Walt.
Damp places. Webster: in Upper and Long Glades.
Preston: in glades near Terra Alta and Reedsville. Fayette:
near Nuttallburg, in sphagnous bog—L. W. N. Monongalia:
at Camp Eden.

L usitatissimum, L. Flax.
An uncommon adventive. Fayette: along the C. &
O. R. R., near Nuttallburg—L. W. N.

GERIANIACÆ.

GERANIUM, L.

G. maculatum, L. Wild Geranium. L. W. N., V, M., M. & G.
Open woods and clearings, frequent throughout the
State. A small form with leaves round in outline and from
1-2 in. in diameter at Bluefield, Mercer county.

G. Robertianum, L. Herb Robert.
Rocks of cool, shaded ravines, rare. Marion: near
Fairmont. Gilmer: near Glenville—V. M.

G. Carolinianum, L. Cranesbill.
Fields, meadows and waste places. Mercer: near Ingleside. Fayette: near Nuttallburg—L. W. N. Kanawha:
near Charleston—Barnes. Monongalia: on the University
Campus; and frequent throughout the State.

FLOERKEA, Willd.

F. proserpinacoides, Willd. False Mermaid.
Wet places. Ohio: near Wheeling—M. & G. Preston: glades near Terra Alta.

OXALIS, L.

O. Acetosella. L. Wood Sorrel.
Deep, rich, mountain woods. Randolph: on Point
Mountain: Cheat Mountain near Cheat Bridge, where this
species grows in such profusion as to actually carpet the
Spruce forests. Grant: near Bayard. Tucker: near Davis;
and Land of Canaan. Gilmer: near Glenville—V. M.

O. violacea, L. M. & G.
Rich, cool woods. Randolph: on Point Mountain.
Monongalia: up Falling Run; at Ullington and Little Falls.
Marion: near Beechwoods; Opekiska and Catawba. Gilmer:
near Glenville—V. M. Fayette: near Nuttallburg, rare—L.
W. N. Hardy: near Moorefield—G. Mercer: near Beaver
Spring.

O CORNICULATA, var. **stricta,** (L.) Sav. Sheep Sorrel. L. W.
N., V. M., M. & G.
Fields, cultivated grounds and roadsides. Common
throughout the State.

O. recurva, Ell.
Open places. Fayette: near Nuttallburg, common—
L. W. N.

IMPATIENS, L.

I. aurea, Muhl.   Pale Touch-me-not.   *I. pallida*, Nutt.   M. & G.
Rich soils near streamlets.   Gilmer: near Glenville—
V. M.   Kanawha: near Charleston.   Fayette: near Nuttallburg—L. W. N.; and common throughout the State.   Also
common in the deep, primitive forests along spring runs,
Randolph: on Rich and Cheat Mountains.   Grant: near
Bayard.   Tucker: along the Blackwater Fork of Cheat.
Hardy: near Moorefield—G.

I. biflora, Walt.   Spotted Touch-me-not.   *I. fulea* Nutt.   L. W. N.,
M. & G.
Shady, moist places, more common than the preceding
species and generally seeking lower altitudes.

RUTACEÆ.

XANTHOXYLUM, L.

X. Americanum, Mill.   Prickly Ash.   Toothache-tree.
Rocky woods, becoming rare.   Jefferson: near Flowing Spring.   Monongalia: Decker's Creek.   Taylor: along
Cheat River.

RUTA, L.

R. GRAVEOLENS, L.   Rue.
Escaped from gardens.   Randolph: on Point Mountain along the road about half way to the summit.

PTELEA, L.

P. trifoliata, L.   Wafer Ash.   Hop-tree.
River banks.   Jefferson: near Harper's Ferry —M. &
G.   Hancock: along Oak Run.   Brooke: on Short Creek.
Summers: near Hinton, on banks of New River, common.

SIMARUBEÆ.

AILANTHUS, Desf.

A. GLANDULOSUS, Desf.   Tree of Heaven.   M. & G.
Naturalized from China.   The seeding-in of this cultivated species is so profuse in the following localities as to
render it a great nuisance.   Monongalia: near Morgantown.
Gilmer: De Kalb P. O.   Lewis: near Weston.   Kanawha:
Pocotaligo.   Jackson: near Sandyville.   Marion: Fairmont.
Gilmer: near Glenville—V. M.   Jefferson: near Harper's
Ferry, and Charlestown.

ILICINEÆ.

ILEX, L.

I. opaca, Ait.　American Holly.
　　Moist woodlands.　Marion: near Nuzums.　Randolph: near Rich Mountains and Laurel Hills.　Kanawha: near Charleston—Barnes.　Fayètte: near Hawk's Nest, large trees 8-12 inches in diameter—Porter; Nuttallburg, in most woods common—L. W. N.　McDowell: near Elkhorn.

I. monticola, Gray.
　　Damp woods.　Randolph: near sumit of Point Mountain; Rich Mountain, near Lone Sugar Knob; Cheat Mountain, near "The Battle Field."　Kanawha: near Charleston—James.

I. mollis, Gray.
　　Deep woods.　Fayette: near Nuttallburg—L. W. N.

I. verticillata, (L.), Gray.　Black Alder.　Winterberry.
　　Frequent in swampy places, throughout the central and northern counties.　Fayette: near Nuttallburg—L. W. N.

CELASTRINEÆ.

EUONYMUS, L.

E. atropurpureus, Jacq.　Burning Bush.　　　　　M. & G.
　　Margins of woods and thickets.　Jackson: near Sandyville.　Gilmer: near Glenville—V. M.　Fayette: near Nuttallburg, rare—L. W. N.

E. Americanus, L.　Strawberry Bush.
　　Rocky, wooded river banks.　Fayette: along the Great Kanawha River, below Gauley Bridge; near Nuttallburg, common—L. W. N.

CELASTRUS, L.

C. scandens, L.　Climbing Bitter-sweet.　Wax-work.　M. & G.
　　Thickets, fence rows and along streams, frequent.　Wood: near Limestone Ridge.　Monongalia and Marion: along the Monongahela River.　Fayette: near Nuttallburg—L. W. N.

RHAMNEÆ.

RHAMNUS, L.

R. Caroliniana. Walt.
McDowell: along Tug Fork of the Big Sandy river near Elkhorn; at Welch, along the same stream.

CEANOTHUS, L.

C. Americanus, L. New Jersey Tea.   M. & G.
Dry, open woods. Upshur: Summit on Staunton Pike. Jackson: Sandyville. Gilmer: Glenville—V. M. Fayette: near Nuttallburg—L. W. N. Hardy: near Moorefield.

AMPELIDEÆ.

VITIS, L.

V. Labrusca, L. Northern Fox-grape.
Damp, rich thickets. Randolph: near Valley Head. Gilmer: near Glenville—V. M. Summers: near Hinton.

V. aestivalis, Michx. Summer Grape.   M. & G.
Thickets. Wirt: along Straight Creek. Randolph: on Point Mountain. Fayette: near Nuttallburg, common— L. W. N. Summers: near Hinton.

V. cordifolia. Michx. Frost Grape.   M. & G.
Thickets and banks of streams. Wirt: along Straight Creek. Randolph: Valley Head; Point Mountain Road. Fayette: near Nuttallburg, common—L. W. N. Summers: near Hinton. Kanawha: near Charleston.

V. riparia, Michx.
Banks of streams. Randolph: near Valley Head; Point Mountain Road. Summers: near Hinton. Jefferson: Shenandoah Je.

V. rupestris, Scheele. Sand Grape. Sugar Grape.
Rocky banks of New river in Fayette near Nuttallburg, plentiful—L. W. N.

V. rotundifolia, Michx. Muscadine. Southern Fox Grape.
Rich river banks. Randolph: near Valley Head. Fayette: near Nuttallburg, rare, on mountain side climbing over trees—L. W. N. Summers: near Hinton.

V. quinquefolia (L.), Lam. Virginia Creeper. American Ivy.
(*Ampelopsis quinquefolia*, Michx.) M. & G.
Woods and thickets. Gilmer: near Glenville—V. M.
Fayette: near Nuttallburg—L. W. N.; and common throughout the State.

CISSUS, L.

C. Ampelopsis, Pers.   (*Vitis indivisa*, Willd.)
River banks. Ohio: near Wheeling—M. & G. Summers: near Hinton.

SAPINDACEÆ.

ÆSCULUS, L.

Ae. glabra, Willd. Ohio or Fetid Buckeye.   M. & G.
Low lands near streams. Wirt: along Straight Creek.
Gilmer: near Glenville—Prof. Brown. Along the Ohio river, common. Monongalia: near Uffington.

Ae. octandra, Marsh.   Sweet Buckeye.   *Ae. flava*, Ait.
Rich mountain woods. Webster: Buffalo Bull Mountains, alt. 2,100 ft. Gilmer: near Glenville—V. M. Summers: near Hinton. Kanawha: near Charleston and Handley. Marion: near Worthington.

Var. **purpurascens**, Gray.
Woodlands. Fayette: near Nuttallburg—L. W. N.

Ae. Pavia, L.
Rich lands along streams. McDowell: along Tug Fork near Elkhorn.

ACER, L.

A. Pennsylvanicum, L.   Striped Maple.   M. & G.
Rich, cool woods. Randolph: on Point Mountain; Staunton Pike on Cheat Mountain. Webster: on Buffalo Bull Mountain. Grant: near Bayard. Tucker: on Blackwater Fork of Cheat. Fayette: near Nuttallburg—L. W. N. and elsewhere in the mountains. McDowell: near Elkhorn.

A. spicatum, Lam.   Mountain Maple.
Same localities as previous species, except Fayette, but more plentiful where found. Greenbrier: near White Sulphur Springs. McDowell: near Elkhorn.

A saccharum, Marsh.　　Sugar Maple.　(*A. saccharinum,*Wang., not L.) V. M., L. W. N., M. & G.
　　　Plentiful throughout the State, especially, however, in the central counties. Randolph: Summit Point Mountain, alt. 3,700 ft. Webster: Buffalo Bull Range, alt. 27–3,600 ft. ·

Var. **nigrum,** (Michx. f.), Britt.　　Black Sugar Maple.　V. M. L. W. N.
　　　With the preceding, almost as plentiful.

A. saccharinum, L.　　White or Silver Maple.　(*A. dasycarpum,* Ehrh.) M. & G.
　　　Banks of Little Kanawha: Gauley River: Great Kanawha, and Ohio. Fayette: near Nuttallburg—L. W. N.; and elsewhere frequent.

A rubrum, L.　　Red or Swamp Maple.　G., L. W. N., M. & G.
　　　Common throughout the State, where it grows upon the hills and in the mountains, as well as in low places.

NEGUNDO, Moench.

N. aceroides, Moench.　　Box Elder, Ash-leaved Maple. M. & G.
　　　Common near rivers throughout the northern and middle counties. Lewis: along Leading Creek. Gilmer: near Glenville—V. M. Summers: near Hinton. Berkeley: near Martinsburg. Hardy: near Moorefield—G.

STAPHYLEA, L.

S. trifolia, L.　Bladder-nut.　　　　　　　　　M. & G.
　　　Rocky woods, thickets and opens. Wirt: near Elizabeth. Monongalia: near Morgantown and Stumptown. Gilmer: near Glenville. Fayette: near Nuttallburg—L. W. N.

ANACARDIACEÆ.

RHUS, L.

R. typhina, L.　Staghorn Sumach.　　　　　　　M. & G.
　　　Dry hillsides. Gilmer: DeKalb Postoffice; near Glenville—V. M. Monongalia: near Stewartown. Summers: near Hinton. Fayette: near Nuttallburg—L. W. N.
　　　During the season of 1890 to 1891 hundreds of plants of this species were noted in Monongalia county to have the inflorescence reverted to leaves, making a strikingly beautiful proliferation when young. This effect, according to Dr. Britton, formed the type of the Linnæan species, *Datisca hirta* (L. Sp. Pl. 1037.) collected by Kalm near Philadelphia, Pa.

R. glabra, L. Smooth Sumach. M. & G.
Frequent or very common in all parts of the State, in rocky or barren soils. Randolph: Point Mountain, alt. 2200 ft.; Cheat Mountains, alt. 27—3600 ft. Cabell: near Barboursville—James. Summers: near Hinton. Fayette: near Nuttallburg, not common—L. W. N. Hardy: near Moorefield—G.

R. copallina, L. Dwarf Sumach. M. & G.
Dry fields and rocky places. Wood: near Lockhart's Run, plentiful. Webster: Buffalo Bull Mountains, alt. 2575 ft. Randolph: Cheat Mountains, alt. 3200 ft. Monongalia: near Ice's Ferry. Fayette: near Nuttallburg, alt. 2000 ft., common—L. W. N.; and frequent throughout the State. Mercer: Beaver Springs and Ingleside.

R. venenata, L. Poison Sumach. Poison Elder.
Swampy places. Randolph: Stalnaker Run. Preston: near Terra Alta, infrequent.

R. radicans, L. Poison Vine. Poison Ivy. Including
R. toxicodendron, L. M. & G.
Thickets and low grounds, very common throughout the State. Monongalia: abundant everywhere in the neighborhood of streams. Webster: Buffalo Bull Mountains, alt. 2100 ft. Kanawha: Pocotaligo. Gilmer: near Glenville—V. M. Fayette: near Nuttallburg—L. W. N. Summers: near Hinton. Mason: near Point Pleasant. Mercer: Princeton, Ingleside and Wills.

R. Canadensis, Marsh. Fragrant Sumach. (_R. aromatica_, Ait.)
Dry or stony soils. Brooke: roadside between Wellsburgh and Bethany College—M. & G. Hardy: near Moorefield—G.

LEGUMINOSÆ.

BAPTISIA, Vent.

B. tinctoria (L.), R. Br. Wild Indigo. "Shoo Fly." M. & G.
Sandy opens. Randolph: along Middle Fork; along Tygart's Valley River; Point and Rich Mountains. Webster: beyond Addison. Nicholas: between long and Collett's Glades. Gilmer: near Glenville—V. M. Kanawha: near Coalburg—James. Fayette: near Nuttallburg—L. W. N. Preston: near Terra Alta.

B. villosa, Ell.
Rocky woods. Mercer: near Ingleside.

B. australis. R. Br.   Blue False Indigo.
River shores.   Ohio: along the Ohio near Wheeling—
M. & G.   Along New river.   Fayette: near Nuttallburg, common—L. W. N.   Summers: near Hinton, abundant.   Mercer: Beaver Spr.

LUPINUS, L.

L. perennis, L.   Lupine.
Sandy soils.   Monongalia: near mouth of Cheat River.

MEDICAGO, L.

M. SATIVA, L.   Lucerne.
Dry Places.   Monongalia: in cinders of railroad near Morgantown, where it has persisted for several years.

M. LUPULINA, L.   Black Medic.
Dry places.   Marion: near Catawba-- K. D. Walker.
Monongalia: near Ufflington.

MELILOTUS, Juss.

M. OFFICINALIS (L.), Lam.   Yellow Melilot.
Ohio: near Wheeling—M. & G.

M. ALBA, L.   White Melilot.   Sweet Clover.   Bokhara Clover.
Roadsides and ditches.   Jackson: near Sandyville.
Wood: near Parkersburg.   Monongalia: near Morgantown.
Berkeley: near Martinsburg.   Jefferson: near Summit Point,
and Shenandoah Junction.   Mason: near Pt. Pleasant.   Mineral: near Keyser.   Hardy: near Moorefield.

TRIFOLIUM, L.

T. ARVENSE, L.   Rabbit-foot Clover.   M. & G.
Established in many places along roadsides and in old
fields.   Kanawha: near Pocotaligo.   Jackson: along C. & P.
Pike.   Mineral: near Keyser—W.   Cabell: near Barboursville—James.   Jefferson: near Charlestown.   Hampshire:
near Romney.

T. PRATENSE, L.   Red Clover.   L. W. N., M. & G.
A common escape to fields, roadsides, and open woods;
even in the higher Alleghenies.

T. REPENS, L.   White Clover.   L. W. N., M. & G.
Fields, open woods, and waste places; common
throughout the State.

T. HYBRIDUM, L.   Alsike Clover.
Becoming frequent in fields and meadows. Monongalia: on the University Campus.

T. AGRARIUM, L.   Yellow Clover.   M. & G.
Sandy hills and roadsides. Upshur: near Buckhannon; summit on Staunton Pike. Randolph: Cheat Mountain Battlefield. Cabell: near Huntington James. Hampshire: near Romney.

T. PROCUMBENS, L.   Low Yellow Clover.   M. & G.
Sandy fields, and roadsides. Kanawha: near Charleston—Barnes. Jackson: plentiful in fields and along roads. Fayette: near Nuttallburg—L. W. N.

TEPHROSIA, Pers.

T. Virginiana, (L.), Pers.   Goat's Rue.
Dry sandy soils. Monongalia: near Morgantown. Gilmer: near Glenville—V. M.

ROBINIA, L.

R. Pseud-Acacia, L.   Yellow Locust. M. & G., V. M., L. W. N.
Common throughout the State, even in the higher mountains.
Dr. Asa Gray, in his account of a "Botanical Excursion to the mountains of North Carolina," says: "On the rocky banks of the Potomac below Harper's Ferry, we saw for the first time the common Locust tree (*Robinia Pseud-acacia*) decidedly indigenous. It probably extends to the southern confines of Pennsylvania; and from this point south, it is everywhere abundant, but we did not meet with it east of the Blue Ridge." The Blue Ridge forms our eastern boundary line between Jefferson county and the State of Virginia. Our State is therefore the eastern extension of this species, though it extends farther north into Pennsylvania.

R. hispida, L.   Bristly or Rose Acacia.   M. & G.
Rich soils. Monongalia: near Morgantown; near Cheat River. Preston: in Laurel Hills. Summers: near Hinton.

ASTRAGALUS, Tourn.

A. Carolinianus, L.   (*A. Canadensis*, L.)   M. & G.
River banks. Monongalia: near Camp Eden. Preston: along Cheat River. Webster: Long Glade. Fayette: near Hawk's Nest—James; near Nuttallburg—L. W. N. Summers: near Hinton.

Specimens found by Mr. Nuttall in his locality resemble so completely Linnaeus' description of A. Caroliniamus—which, however, is not sufficiently different from his A. Canadensis published later, to consider these as two species—the former must, therefore, take the precedence and stand for the species.

A. —————

An odd species widely different from any known eastern form, which cannot be named on account of lack of fruit on the specimens collected, and our unaccountable inability to find any on the plants later in the season.

Hardy: near Moorefield—A. D. Hopkins.

STYLOSANTHES, Swartz.

S. biflora (L.), B. S. P.   Pencil Flower (*S. elatior*, Sw.)   M. & G.
Dry, open woods. Wirt: near Burning Springs. Upshur: summit on Staunton Pike. Summers: near Hinton. Fayette: near Nuttallburg, rocky banks of New River—L. W. N. Monongalia: near Camp Eden.

S. hamata (L.), Britt.   (*S. procumbens*, Siv.)
Shores of New River. Summers: near Hinton. First report of this species north of Tennessee.

DESMODIUM, Desf.

D. nudiflorum, (L.), DC.   Tick Tree-foil.   M. & G.
Rich woods, common. Wood: near Leachtown. Randolph: on Point Mountain. Webster: Buffalo Bull Mountain. Gilmer: near Glenville—Prof. Brown. Fayette: near Nuttallburg—L. W. N. Summers: near Hinton.

D. grandiflorum. (Walt.), DC.   *D. acuminatum*, D. C.
Rich woods. Monongalia: Marion: Preston: Wetzel: Mineral: Jefferson: Berkeley and Calhoun counties. Fayette: near Kanawha Falls—James; near Nuttallburg—L. W. N. Summers: near Hinton. Kanawha: near Charleston; and frequent throughout the State.

D. rotundifolium, (Michx.), DC.   "Hive Vine."
Dry, rocky woods. Monongalia: near Morgantown. Lewis: along Leading Creek. Upshur: near Lawrence. Fayette: near Nuttallburg, alt. 1500 ft.—L. W. N.

D. ochroleuca, M. A. Curtiss.   M. & G.
Mineral: along Knobby Mountains. Jefferson: near Millville.

D. canescens (L.), DC.
> Open woods and clearings. Wood: near Lockhart's Run. Monongalia: campus, Morgantown. Summers: Riffe; Wolf Creek. An abundant weed. Fayette: near Nuttallburg—L. W. N. Monroe: near Alderson; the worst weed in some fields. Mason: near Point Pleasant.

D. cuspidatum (Muhl.), Hook.
> Thickets. Monongalia: along Decker's Creek; near the mouth of Cheat River. Marion: near Little Falls. Wood: near Kanawha Station. Fayette: near Nuttallburg —L. W. N. Summers: near Hinton.

D. Dillenii, Darl.
> Open woodlands. Monongalia, Wood, Marion, Wirt, Upshur, Lewis, Jefferson, Berkeley, Grant and Preston counties. Monroe: near Alderson.

D. paniculatum, (Nutt.), DC.
> Copses. Wood, Wirt, Calhoun and Gilmer, along the Little Kanawha River. Monongalia and Marion: along the Monongahela River. Fayette: near Nuttallburg—L. W. N. Summers: near Hinton.

D. Canadense, (L.), DC.   M. & G.
> Dry, but rich woodlands. Monongalia: at The Flats. Marion: near Opekiska. Wood: near Kanawha Station. Mason: Point Pleasant.

D. rigidum, (Ell.), DC.
> Dry hillsides. Monongalia: near Morgantown; on Dorsey's Knob. Mineral: along Knobby Mts. Marion: opposite Montana.

D. ciliare, (Muhl.), DC.
> Dry hillsides. Monongalia: Cheat View, Little Falls Beechwoods. Marion: near Houghtown. Lewis: near Weston. Upshur: near Buckhannon.

D. Marilandicum, (L.), Boott.
> Copses. Grant: near Bayard. Mineral: near Keyser. Berkeley: North Mountain. Morgan: near Hancock. Jefferson: Shenahdoah Jc. Summers: near Hinton.

D. lineatum, (Michx.), DC.
> Dry soils. Jefferson: near Shenandale Springs. Gilmer: near Glenville—V. M.

351

LESPEDEZA, Michx,

L. repens, (L.). Bart.   Bush Clover.   (*L. procumbens*, Michx.
Dry, sandy soils.   Monongalia: hills near The Flats;
banks of the Monongahela below Dille's.   Wood: near Kanawha Station.   Kanawha: Charleston—James.   Summers:
near Hinton.

L. reticulata, (Muhl.), Pers.
River banks.   Fayette: near Nuttallburg, along New
River—L. W. N.

L. violacea, (L.), Pers.
Dry copses.   Monongalia: near Morgantown.   Wood:
near Parkersburg.   Wirt: near Burning Springs.   Fayette:
near Nuttallburg—L. W. N.   Summers: near Hinton.

L. Virginica, (L.), Britt.   (*L. sessiliflora*, Michx.)
River shores.   Summers: near Hinton, along New
River.

L. Stuvei, Nutt.
Mountain sides.   Fayette: near Nuttallburg—L.
W. N.

Var. **intermedia,** Watson.
River banks.   Fayette: near Nuttallburg, along New
River, plentiful—L. W. N.

L. polystachya, Michx.
Dry opens.   Fayette: near Nuttallburg, alt. 2000 ft.
L. W. N.

L. capitata, Michx.
Dry sandy soils.   Ohio River banks, frequent.   Mon-
ongalia: near Morgantown.   Marion: near Catawba, and
Houghtown.

L. STRIATA, (Thunb.), Hook. & Arn.   Japanese Clover.
Dry, red soils.   Spreading profusely along the C. & O.
R. R. in Kanawha, Putnam and Cabell counties.

VICIA, L.

V. Caroliniana, Walt.   Carolina Vetch or Tare.
River banks and edges of glades.   Webster: Welsh
Glade; island in Long Glade.   Wood: shores of Little Kan-
awha River.   Gilmer: near Glenville—V. M.   Fayette: near
Nuttallburg—L. W. N.   Monongalia: near Cassville.

LATHYRUS, L.

L. venosus, Muhl.
Shady banks. Mineral: banks of the Potomac near Keyser. Fayette: near Nuttallburg, a form 10–15 flowered, with winged stem and stipules 3–15 lines long—L. W. N.

AMPHICARPÆA, Ell.

A. comosa (L.), Ridd. *A. monoica, Ell.*
Rich, damp thickets. Monongalia and Marion: along the Monongahela River. Fayette: near Nuttallburg—L. W. N.

APIOS, Moench.

A. tuberosa, Moench. Ground Nut.
Low Grounds along streams. Frequent throughout the State. Fayette: near Nuttallburg, not common—L. W. N.

PHASEOLUS, L.

P. polystachus, (L.), B. S. P. Wild Kidney Bean. *P. perennis*, Walt.
Copses. Wood: near Kanawha Station. Monongalia: near Morgantown. Mason: near Point Pleasant.

P. helvolus, L. *Strophostyles angulosa*, Ell.
Sandy river banks. Mason: banks of the Ohio near Point Pleasant. Fayette: near Nuttallburg, with the inner surface of the petals pink—L. W. N.

GLEDITSCHIA, L.

G. triacanthos, L. Honey or Black Locust. M. & G.
Rich woods. Monongalia: near Morgantown. Wirt: on Nigh-Cut Hill. Randolph: Rich Mts.; Point Mt., alt. 233,700 ft. Gilmer: near Glenville. Hardy: near Moorefield.

CASSIA, L.

C Marilandica, L. Wild Senna. M. & G.
Sandy alluvium. Fayette: near Nuttallburg—L. W. N.; near Gauley Bridge. Gilmer: near Glenville—Prof. Brown. Kanawha: near Pocotaligo. Jackson: near Sandyville. Jefferson: near Harpers Ferry. Mason: near Point Pleasant. Harrison: near Shinnston. Summers: near Hinton. Monongalia: near Morgantown. Hardy: near Moorefield G.

C. Chamæchrista, L.   Partridge Pea.   M. & G.
Sandy fields. Monongalia: near the mouth of Cheat River. Marion: near Opekiska. Fayette: near Nuttallburg L. W. N.

C. nictitans, L.   Wild Sensitive Plant.   M. & G.
Sandy places. Monongalia: along the Monongahela River. Marion: near Clements. Fayette: near Nuttallburg—L. W. N. Mason: near Point Pleasant. Summers: near Hinton. Hardy: near Moorefield—G.

CERCIS, L.

C. Canadensis, L.   Red-bud. Judas' Tree.   M. & G.
Especially common on hillsides along the Great Kanawha River in Putnam and Mason counties. Monongalia: along Cheat River; and the Monongahela. Wirt: along Little Kanawha River. Gilmer: near Glenville—V. M. Fayette: near Nuttallburg—L. W. N. Summers: near Hinton.

GYMNOCLADUS, Lam.

G. dioicus, (L.), Koch. Kentucky Coffee-tree. *G. Canadensis,* Lam.
M. & G.
Rich woods, infrequent. Randolph: Point Mountain, beyond Valley Bend. Webster: Buffalo Bull Mountain, along ridge.

ROSACEÆ.

PRUNUS, L.

P Americana. Marsh.   Wild Yellow or Red Plum.   M. & .G
River banks and woodlands. Monongalia: near Morgantown. Marion: near Opekiska. Tyler: near Long Reach —Col. Johnson. Wood: near Lockhart's Run. Gilmer: near Glenville—V. M.

P. Chicasa, Michx.
Thickets. Monongalia: Permian formations near Cassville. Tyler: near Long Reach.

P. Pennsylvanica, L. f.   Wild Red Cherry.   V. M., M. & G.
Rocky woods. Very common throughout eastern portion of the State.

P. Virginiana, L.   Choke Cherry.   V. M., M. & G.
Moist, shady places, common.

P. serotina. Ehrh. Wild Black Cherry. G., L. W. X., V. M.,M.& G.
Common throughout the State, where it often forms
extensive and very valuable forests. This is especially true
of the tree in the central eastern section.

PHYSOCARPA, Raf.

P. opulifolia (L), Raf. Nine-bark.
Along streams, common. Monongalia: along the Mo-
nongahela river. Webster: Upper Glade. Fayette: near
Nuttallburg—L. W. X. Hardy: near Moorefield.

SPIRAEA, L.

S. betulaefolia, Pall. Birch-leaved Meadow-sweet.
Rich mountain woods. Webster: near Upper Glade.
Hardy. near Moorefield--G.

S. Virginiana, Britt. West Virginia Meadow-sweet.
Discovered 1890 along the Monongahela river near
Morgantown. The following description is taken from Prof.
Britton's account of the plant in "Bull. Torrey Club," Dec.
1890:

"A glabrous shrub, the branches forming long wands,
erect or reclining, 1-4 ft. long. Leaves oblong or slightly ob-
lanceolate, thin, obtuse or short-pointed at the apex, rounded
or cuneate at the base, 1½ to 2 in. long, 5-8 lines wide, green
above, pale beneath, entire or with a few low serrations in
the upper half; petioles 2 lines long: pedicels and peduncles
pale and glaucous: flowers about 2-lines broad, in terminal
compound corymbs 1-3 in. across; calyx teeth 5, triangular,
blunt, about the length of the short-campanulate tube, dis-
tinctly glaucous: petals 5, white, ovate-orbicular, obtuse;
stamens 15-20, persistent; styles 5-6; follicles in the specimens examined, 5-6, apparently sterile, included in the persistent calyx."

"On damp rocksalong the Monongahela river, Morgantown. West Virginia, collected by Dr. C. F. Millspaugh in
flower. June 20th, 1890, and in apparently imperfect fruit
late in September. Collected also by Mr. G. R. Vasey in the
mountains of North Carolina, 1878."

"*Spiraea betulaefolia*, Pall. and *S. corymbosa*, Raf., have
much longer follicles exerted beyond the calyx, broader,
thicker, and dentate leaves, and are different in habit. Ra-
finesque published a number of species in his New Flora, but
none of them can apply to this one."

S. tomentosa, L. Hardhack. Steeple-bush. M. & G.
Low grounds. Webster: Welch Glade. Wood:
near Lockhart's Run.

S. rubra, (Mill.) Britt.   Queen of the Prairie.   *S. lobata*, Jacq.
Meadows.   Monongalia: near Morgantown.   Preston:
near Terra Alta.

S. Aruncus, L.   Goat's Beard.
Rich ground and along streams.   Monongalia: near
Morgantown, Uffington, Little Falls, Day Creek and Gran-
ville along the Monongahela.   Randolph: along Middle
Fork, on Rich Mountains, alt. 2.125 ft.   Gilmer: near Glen-
ville—V. M.

GILLENIA, Moench.

G trifoliata, (L.), Moench.   Bowman's Root.   Indian Physic.
M. & G.
Rich woods, frequent.   Webster: Welch and Long
Glades.   Monongalia: along the Monongahela River.   Little
Falls to Beech Woods.   Mineral: near Keyser – W.   Sum-
mers: near Hinton.   Hardy: near Moorefield—G.   Mercer:
near Ingleside.

G stipulacea, (Pursh), Nutt.   American Ipecac.
Borders of woods.   Wood: between Kanawha Station
and Lockhart's Run. .

RUBUS, L.

R. odoratus, L.   Purple-flowering Raspberry.   Thimble-berry.
M. & G.
Damp rocky places.   Monongalia: near Little Falls.
Grant: near Bayard.   Randolph: near Crickard P. O.   Fayette:
near Kanawha Falls—James; near Nuttallburg—L. W. N.
Summers: near Hinton.   Hardy: near Moorefield—G.

Var. **Columbianus.**   Columbian Raspberry.
Leaves ample, 5-7-incised, divisions oblong-lanceolate
long and taper pointed, sharply and mostly double serrate.
Inflorescence smaller and more compact.   Fruit smaller than
in the species and of a more decided musky taste.   Monongalia: cool woods, Tibbs Run.

R. strigosus, Michx.   Wild Red Raspberry.   M. & G.
Thickets.   Fayette: near Nuttallburg, not plentiful
—L. W. N.   Pocahontas: Spruce Mountain—A. D. Hopkins.

R. occidentalis, L.   Black Raspberry.   V. M., M. & G., L. W. N.
Frequent throughout the State.

R. villosus, Ait.   High Blackberry.   L. W. N., V. M., M. & G.
Common everywhere in the State.

Var. **humifusus,** T. *&.* G.
 Woods and river banks. Fayette: near Nuttallburg
—L. W. N.

Var. **frondosus,** Torr.
 Fayette: near Nuttallburg— L. W. N. Preston: near
Tunnellton.

R. **Millspaughi,** Britt.
 This species was described in "The Bulletin of the
Torry Club" for 1891, page 366, as follows:
 "Ascending, wand-like, entirely unarmed or with a
"very few weak prickles above, glabrous throughout or the
"younger shoots scurfy pubescent. Stems one and one-half
"to four meters long; leaves long petioled, pedately 5-foliate
"or some of those on the twigs 3-foliate; leaflets thin, oval,
"glabrous on both sides, long-acuminate at the apex, mostly
"rounded at the base, 12-15 cm. long, about 5 cm. wide,
"sharply, but not deeply serrate; stock of the terminal leaf-
"let 7-10 cm. long; inflorescence loosely racemose; bracts
"linear-lanceolate; acuminate; fruit black, about 10 mm.
"long."
 "Nearest to R. villosus, but evidently a distinct species. Curiously enough there is a leaf of this species glued
down on the sheet of R. Canadensis, L in herb Linn., and it
appears to have been included in his description of that species—the specimens furnished by Kalm."
 Near the summit of Point Mountain in Randolph
county at an altitude of 3,500 ft., also along the Gandy in
great profusion. Pendleton and Pocahontas: on Little Rich
Mountains abundant. The mountaineers claim that it is upon
this species that the bears grow fat for their period of hiber-
nation, the fruit being late to ripen and very nutritious.

R. **Canadensis,** L. Dewberry. L. W. N., V. M., M. & G.
 Abundant on dry hillsides throughout the State.

Var. **roribaccus,** Bailey. Leucretia Dewberry.
 Dry hillsides. Randolph: near Beverly.
 This new variety of the species is described by Prof.
Bailey in the American Garden, November, 1890, as follows:
"Plant larger and stronger; leaflets broad below, usually tri-
angular-ovate, doubly serrate with small teeth, and more or
less notched and jagged; peduncles longer, straighter and
stouter, habitually more numerous and more conspicuously
overtopping the leaves; flowers very large (some times two
inches across); sepals uniformly larger, some of them much
prolonged and leaf-like and conspicuously lobed (some times
becoming an inch long and wide); fruit much larger."

R. hispidus, L. Running Swamp-Blackberry. M. & G.
Glade regions of Preston, Randolph and Webster counties. Fayette: near Nuttallburg—L. W. N.

R. trivalis, Michx. Low Bush-blackberry.
Sandy soil. Kanawha: near Charleston—James.

DALIBARDA, L.

D. repens, L.
Deep mountain woods. Grant: near Bayard. Tucker: along Blackwater Fork of Cheat.

GEUM, L.

G. Canadense, Jacq. (*G. album*, Gmel. M. & G.
Woods and thickets. Calhoun, Wood, Grant, Mineral, and Monongalia counties. Fayette: near Kanawha Falls— James; near Nuttallburg—L. W. N. Summers: near Hinton.

G. Virginianum, L.
Borders of woods and low grounds. Wood, Wirt, Calhoun, Gilmer, Marion, Lewis, Monongalia, and Jefferson counties.

G. vernum (Raf.), T. & G.
Moist places. Monongalia: near Morgantown; Little Falls. Marion: near Opekiska.

WALDSTEINIA, Willd.

W. fragarioides, (Michx.), Tratt. Barren Strawberry.
Wooded hillsides. Grant: near Bayard.

FRAGARIA, L.

F. Virginiana, Mill. Wild Strawberry. M. & G.
Moist woodlands and fields, common. Monongalia: near Morgantown—M. H. Brown. Gilmer: near Glenville —V. M. Fayette: near Nuttallburg—L. W. N. Hardy: near Moorefield—G.

F. vesca, L. Northern Wild Strawberry. M. & G.
Shady woods, less common than the last. Kanawha: near Charleston—James. Fayette: near Nuttallburg—L. W. N.

F. INDICA, Andr. Indian Strawberry.
Escaped to waste places. Monongalia: near Morgantown. Kanawha: near Charleston—Barnes.

POTENTILLA, L.

P. Norvegica, L.   M. & G.
Fields and wet places. Wood: near Kanawha Station. Monongalia: near Morgantown. Grant: near Bayard. Fayette: near Nuttallburg, rare—L. W. X. Tucker—near Davis.

P. Canadensis, L.   Cinquefoil.   G., L. W. X., V. M., M. & G.
Dry fields, an abundant weed throughout the State.

Var. **simplex** (Michx.), T. & G.
Dry fields. Ohio: near Wheeling—M. & G.

AGRIMONIA, L.

A. Eupatoria, L.   L. W. X., M. & G.
Borders of woods, frequent throughout the State.

A. parviflora, Ait.
Woods and glades. Randolph: on Lone Sugar Knob. Preston: near Terra Alta. Webster: Long Glade, Nicholas: Collett's Glade. Tucker: near Davis. Fayette: near Nuttallburg, alt. 2,000 ft.—L. W. X. Summers: near Hinton. Monroe: near Alderson.

POTERIUM, L.

P. Canadense (L.), Gray. Burnet.
Rich, moist woods. Randolph: along Cheat river. Tucker: along Blackwater Fork of Cheat. Monongalia: Cheat river near Camp Eden. Preston: Terra Alta.

ROSA, L.

R. Carolina, L.   Carolina Rose.   M. & G.
Damp places. Wood: near Kanawha Station, profuse. Upshur: near Buckhannon. Webster: Long Glade. Randolph: meadows along Tygart's Valley river.

R. humilis, Marsh.   Dwarf Wild Rose.
Dry soils. Wood: near Kanawha Station. Monongalia: near Uffington and Beechwoods. Gilmer: near Glenville—V. M. Fayette: near Nuttallburg—L. W. X. Summers: near Hinton.

Var. **lucida** (Ehrh.), Best.   Larger Wild Rose.   (*R. lucida*, Ehrh.)
                                                                M. & G.
Shaded hillsides. Kanawha: near Charleston—James – Barnes. Monongalia: plentiful along Cheat river above Camp Eden. Fayette: Kanawha Falls—James.

R. RUBIGINOSA, L. Sweet-brier. Elgantine. M. & G.
Frequent along roadsides and thickets. Nicholas:
along Gauley river. Randolph: Staunton Pike on Cheat
Mountains. Summers: near Hinton. Monongalia: near
Lee's Ferry.

R. CANINA, L.
Rocky Banks. Fayette: near Nuttallburg—L. W. N.

PYRUS, L.

P. coronaria, L. Wild Crab Apple. M. & G.
Opens and damp places. Monongalia: frequent about
Morgantown. Marion: along the Monongahela river. Gil-
mer: near Glenville—V. M. Fayette: near Nuttallburg—
L. W..N. Mercer: near Beaver Spr.

P. angustifolia, Ait. Narrow-leaved Crab.
Glady regions of Preston, Webster and Nicholas counties.

P. arbutifolia (L.), L. f. Choke Berry.
Damp places. Webster: Upper Glade. Preston: near
Terra Alta. Nicholas: Collett's Glade. Fayette: near Nut-
tallburg, alt. 2000 ft.—L. W. N.

Var. **melanocarpa** (Willd.), Hook.
Preston: Reedsville Glade; Morgans' Glade. Webster:
Upper and Welch Glades.

P. Americana, (Marsh.) DC. Mountain Ash. M. & G.
Damp mountain woods. Randolph: near Cheat
Bridge. Grant: near Bayard. Tucker: near Davis; and
along the Blackwater.

CRATEGUS, L.

C. spathulata, Michx.
Rocky Woods. Mercer: near Beaver Spring, and
Wills. McDowell: near Elkhorn.

C. cordata, Ait.
Rocky mountain woods. Mercer: near Beaver Spring,
Ada, and Ingleside.

C. OXYACANTHA, L.
River banks. Fayette: near Nuttallburg—L. W. N.

C. apiifolia, Michx.
Rocky woods. Mercer: near Ingleside and at Beaver
Spring.

C. coccinea, L.   Scarlet Thorn. " ,   M. & G.
Thickets. Calhoun: Lower Leading Creek. Monongalia: near Ice's Ferry. Marion: near Opekiska. Gilmer: near Glenville—V. M. Fayette: near Nuttallburg - L. W. N. Upshur: near Buckhannon. Summers: near Hinton. Hardy: near Moorefield—G. Mercer: near Ingleside.

C. mollis (T. & G.), Sarg.   *C. coccinea*, var. *mollis*, T. & G.
Mercer: near Beaver Spr. and Ingleside.

C. tomentosa, L.   Black Thorn.   M. & G.
Uplands. Monongalia: near Stewartstown and Uffington; road to Dorsey's Knob. Randolph: near Cheat Bridge. Tucker: along the Blackwater. Wirt: near Elizabeth. Wood: near Lockhart's Run. Summers: near Hinton. McDowell: near Elkhorn.

C. punctata, Jacq.   M. & G.
Borders and open woods. Monongalia: near Ice's Ferry; Cheat River above Camp Eden. Greenbrier: near White Sulphur Springs. Fayette: near Nuttallburg—L. W. N.

C. Crus-galli, L.   Cockspur Thorn.
Thickets. Monongalia: frequent. Marion: near Houghtown. Preston: near Reedsville and Terra Alta Mineral: near Keyser. Jefferson: near Shenandoah Junction. Upshur: near Buckhannon. Fayette: near Nuttallburg—L. W. N. Greenbrier: near White Sulphur Springs. Summers: near Hinton. Hardy: near Moorefield—G. Mercer: near Beaver Springs and Ingleside.

C. flexispina, (Moench.) Sarg.   Summer Haw.   *C. flava*, Ai
Shady river banks. Fayette: banks of New River near Nuttallburg—L. W. N. Mercer: near Ingleside. M Dowell: along Tug Fork, near Welch.   •

Var. **pubescens** (Gray).
Sandy woods. Mercer: along stream opposite Will

C. uniflora, Moench.   *C. parviflora*, A
Sandy woods. Mercer: along streamlet near Ing side.

AMELANCHIER, Lindl.

A. Canadensis, (L.), Medic.   Shad Bush. June Berry.
W. N. V. M., M. & G.
Common generally, even in the higher mountains.

CALYCANTHEÆ.

BEURERA, Ehret.

(Calycanthus, L.)

B florida (L.). Allspice. Sweet-scented Shrub.
Rich woods. Randolph: near Fords; on Staunton
Pike, thence frequent over Rich Mountains. Webster and
Nicholas: Along Buffalo Bull Ridge. Fayette: along the
Gauley river near Gauley Mountains. Summers: near Hinton.

B. laevigatus, (Willd.)
Rich woods. McDowell: back of R. R. water tank
near Welsh.

SAXIFRAGEÆ.

ASTILBE, Don.

A. decandra, Don. Goats-beard.
Rich woods. Mercer: near Wills.

SAXIFRAGA, L.

S. Virginiensis, Michx. Early Saxifrage. L. W. N., V. M., M.
& G., G.
Exposed moist rocks and opens. General throughout
the State.

S. Pennsylvanica, L. Swamp Saxifrage.
Glades. Preston: Morgan's and Terra Alta Glades.

S. micranthifolia (Haw.). Lettuce Saxifrage. *S. erosa,* Pursh.
M. & G.
Spring rills in deep woods. Monongalia: near Camp
Eden. Grant: near Bayard. Tucker: along the Blackwater.

THEROFON, Raf.

T. aconitifolia (Nutt.). *(Boykinia aconitifolia, Nutt.)*
Creek beds. Fayette: near Nuttallburg; plentiful
along the beds of creeks at an alt. of 1800 ft., does not extend
down these beds as far as the shores of New river (alt. 1100
ft.). Often 3-celled, sometimes the flowers 6-parted, few
7-parted—L. W. N.

TIARELLA, L.

T. cordifolia, L. False Mitre-wort.
Rich, moist woods. Common throughout the Alleghanies and the foot-hills. Monongalia: near Morgantown. Gilmer: near Glenville—V. M. Fayette: near Nuttallburg — L. W. N. Greenbrier: near White Sulphur Springs. McDowell: near Elkhorn. Mercer: near Bluestone Je.

MITELLA, L.

M. diphylla, L. Mitre-wort. Bishop's Cap.   L. W. N., V. M.,
M. & G.
Rich, shady woods. Common throughout the State.

HEUCHERA, L.

H. villosa, Michx.
Rocky places. Fayette: Kanawha Falls—James; Loup Creek—James. Nuttallburg- L. W. N. Kanawha: near Coalburg—James.

H. Americana, L. Alum-root.   M. & G.
Rich, damp woods. Monongalia: near Morgantown, and frequent elsewhere. Gilmer: near Glenville—V. M. Fayette: near Nuttallburg—L. W. N. Greenbrier: near White Sulphur Springs. Grant: near Bayard. Tucker: near Davis. Hampshire: near Romney.

H. Rugelii, Shuttlw.
Shaded cliffs. Fayette: near Nuttallburg, common --L. W. N.

PARNASSIA, L.

P. grandiflora, DC.   *P. Caroliniana,* Michx.
Wet banks. Fayette: near Kanawha Falls—Selby.

HYDRANGEA, L.

H. arborescens, L. Wild Hydrangea.   M. & G.
Rich opens. Monongalia: near Morgantown. Marion: Opekiska. Wood: near Kanawha Station. Wirt: near Burning Springs. Gilmer: near Glenville—V. M. Lewis: along Stone Coal Creek. Throughout the above range the flowers were all fertile. Fayette: near Hawk's Nest —James; near Kanawha Falls- James; near Nuttallburg— L. W. N.
A form with grass-green marginal radiant flowers, in a deep ravine in Fayette: near Nuttallburg.

Var. **Kanawhana.**
Low straggling bush, leaves small, paler beneath, acuminate, somewhat cordate at the base; cymes very open and branching, marginal radiant flowers many, 1 in. broad, fertile flowers nearly glabrous, smaller than in the species. Along the Little Kanawha River from Kanawha Station to Glenville.

RIBES, L.

R. Cynosbati, L.   Prickly Gooseberry.   M. & G.
Deep rocky woods. Randolph: along Cheat River, alt. 3,360 ft.; Point Mountain, alt. 3,700 ft.; Rich Mountain, alt. 2,700. Grant: near Bayard. Preston: near Terra Alta; and frequent throughout the northern and eastern counties. Hardy: near Moorefield.

R. rotundifolium, Michx.   Smooth Gooseberry.
Rich, cool, mountain woods, with the last, frequent.

R. prostratum, L., Her.
Pocahontas: summit of Spruce Knob, alt. 4800 ft.—A. D. Hopkins.

R. floridum, L'Her.   Wild Black Currant.
Rich woods. Randolph: near Beverly. Grant: near Bayard. Preston: near Terra Alta. Fayette: near Nuttallburg—L. W. N.

CRASSULACEÆ.

SEDUM, L.

S. pulchellum, Michx.
Rocky places. Jefferson: near Harper's Ferry—Asa Gray.

S. Nevii, Gray.
Dry, rocky places. Greenbrier: near White Sulphur Springs.

S. ternatum (Haw), Michx.   Stone-crop.   G., L. W. N., V. M., M. & G.
On rocks in deep woods and opens. Throughout the State.

S. telephioides, Michx.   M. & G.
Drier situations. Throughout the northern counties. Jefferson: Harper's Ferry—Asa Gray. Hardy: near Moorefield—G. Hampshire: near Romney.

S. TELEPHIUM, L. Live-for-ever.
Along railroad banks. Jefferson, Morgan and Berkeley counties.

PENTHORUM, L.

P. sedoides, L. Ditch Stone-crop.   L. W. N., M. & G.
Open wet places, and ditches. Throughout the State.

DROSERACEÆ.

DROSERA, L.

D. rotundifolia, L. Sundew.   M. & G.
Glades. Preston: Cranberry Summit; Morgan's Glade; and Terra Alta.

HAMAMELIDEÆ.

HAMAMELIS, L.

H. Virginica, L. Witch-hazel.   L. W. N., V. M., M. & G.
Common in damp woods throughout the State.

LIQUIDAMBAR, L.

L styraciflua, L. Sweet-gum. Alligator-wood.   M. & G.
Rich woods. The distribution of this species in the State according to my notes of travel is, from east to west, as follows: Beginning near the source of Peter Creek. in Nicholas county, it follows that stream to its junction with the Gauley river. down this to its confluence with the New river to form the Great Kanawha, which latter it follows to the mouth of Elk river, whence it bears northward up Eight Mile creek to the Pocotaligo and its Middle Fork, across to Mill Creek in Jackson, which it follows to the Ohio.
It is also noted in Gilmer: near Glenville—V. M. Fayette: near Kanawha Falls—James; near Nuttallburg—L. W. N. Cabell: near Huntington—Selby. I have also met with it in Summers: near Hinton; and along the Greenbrier river in that county.

HALORAGEÆ.

CALLITRICHE, L.

C. heterophylla, Pursh.
Fayette: near Nuttallburg—L. W. N.

MELASTOMACEÆ.

RHEXIA, L.

R. Virginica, L. Meadow Beauty.
Moist, sandy meadows, and river shores. Monongalia: near Camp Eden; Little Falls. Wood: near Lockhart's Run. Wirt: near Burning Springs. Upshur: near Lorentz. Randolph: along Tygart's Valley River. Berkeley: near Martinsburg. Putnam: near Buffalo.

LYTHRARIÆ.

CUPHÆA, R. Br.

C. petiolata, (L.), Koehne. "Tar-weed." *C. viscosissima*, Jacq.
Dry soils, and fields. Monongalia, Marion, Wood, Wirt, Calhoun, Gilmer, Lewis, Upshur and Randolph. Fayette: near Gauley Bridge; near Nuttallburg—L. W. N.; near Hawk's Nest—James. Greenbrier: near White Sulphur Springs. Summers: near Hinton. Monroe: near Alderson. Berkeley: near Martinsburg, and elsewhere.

ONAGRARIEÆ.

EPILOBIUM, L

E. spicatum, Lam. Fire-weed. *E. angustifolium*, L. M. & G.
In new clearings. Mineral: Grant: and Tucker: along the W. Va. Cent. R. R. Randolph: summit of Point Mountain, alt. 3,700 ft. Cheat Mountain, alt. 27-3600 ft. Hardy: near Moorefield—G.

E. coloratum, Muhl. Willow-herb. M. & G.
Ditches, and wet rocks. Greenbrier: near White Sulphur Springs. Fayette: near Nuttallburg—L. W. N. Mason: near Pt. Pleasant; and frequent throughout the State.

LUDWEGIA. L.

L. alternifolia, L. Seed-box. M. & G.
Wet banks. Wood: near Lockhart's Run. Monongalia: along Cheat River near Camp Eden. Fayette: near Kanawha Falls—James; near Nuttallburg—L. W. N.

Var. **linearifolia**, Britt. n. var.
With or near the species. Wood: near Lockhart's Run. Fayette: near Nuttallburg—L. W. N.
Described by Prof. Britton in "Bull. Torrey Club," Dec., 1890, as follows.

"Two or three feet high, divergently branched, the branches ascending. Leaves linear, elongated. 2-4-in. long. 1½-4-lines wide, acute; flowers solitary in the axils of the upper leaves or bracts, yellow; sepals ovate-lanceolate acute, narrower than those of L. alternifolia; branches and both sides of the leaves somewhat pubescent. Petals apparently remaining on the plant longer than those of L. alternifolia, which, as Dr. Millspaugh observes, commonly fall away when the plant is shocked."

"Appearing very distinct from typical L. alternifolia, but presumably but a variety of it. From the description it may be the *Rhexia linearifolia*, Poir. in Lam. Encycl. vi. 2, said to come from Carolina."

L palustris (L.), Ell.

Sandy soil. Fayette: in a Sand bar in New river near Nuttallburg—L. W. N.

OENOTHERA, L.

Oe. biennis, L.   Evening Primrose.   L. W. N., M. & G.

Frequent or common. throughout the State.

Var. **grandiflora** (Ait.), Lindl.

Frequent. Randolph: Cricard, P. O.; Point Mountain. Wood: near Kanawha Station. Preston: near Tunnellton; near Terra Alta. Greenbrier: near White Sulphur Springs. Hardy: near Moorefield—G.

Oe. pumila. L.

Dry fields, frequent throughout the State, especially in the northern section. Hardy: near Moorefield—G.

Oe. fruticosa, L.   Sun-drops. "Wild Beet."   M. & G.

Common in most soils, and in cultivated fields as a weed. Summers: near Talcott and Lowell. Marion: near Worthington. Gilmer: near Glenville—V. M. Fayette: near Nuttallburg—L. W. N.

Var. **linearis** (Michx.), Watson.

Damp places. Wood: near Kanawha Station. Wirt: near Elizabeth. Calhoun: near Grantsville. Gilmer: near Glenville—V. M. Upshur: near Buckhannon.

Var. **differta**. Crowded Sun-drops.

Damp meadows. Wood: near Lockhart's Run, the most common form.

Stems 1-2-ft. high, nearly smooth, branching diffusely from every axil. Flowers profuse, large. Lower leaves ovate. Capsules narrowly winged, very short; apical inflorescence strongly cymose.

GAURA, L.

G. biennis, L.   M. & G.
Dry banks. Webster: near Taylor. Greenbrier: near White Sulphur Springs. Fayette: near Nuttallburg—L. W. N. Harrison: near Lumberport. Berkeley: near Martinsburg. Hardy: near Moorefield—G.

CIRCÆA. L.

C. Lutetiana, L. Enchanter's Nightshade.   M. & G.
Low grounds, and wet woods. Lewis: up Stone Coal Creek. Gilmer: near Glenville—Prof. Brown. Fayette: near Nuttallburg—L. W. N.

C. alpina, L.   M. & G.
Deep, rich woods. Randolph, Grant, Tucker and Pendleton: prevalent in the Alleghanies. Gilmer: near Glenville—V. M. Monongalia: along Cheat river, above Camp Eden.

PASSIFLOREÆ.

PASSIFLORA, L.

P. lutea, L. Yellow Passion-flower.
Hillsides. Fayette: near Nuttallburg—L. W. N.

P. incarnata, L. Passion-flower.
Dry soil. Gilmer: near Glenville—V. M.

CUCURBITACEÆ.

CUCURBITA, L.

C. OVIFERA, L. Gourd.
Escaped to waste grounds. Monongalia: about Morgantown. Jefferson: near Shepherdstown.

CITRULLUS, L.

C. VULGARIS, Schrad. Watermelon.
Escaped to waste grounds. Mason: banks of the Ohio near Point Pleasant.

CUCUMIS, L.

C. MELO, L. Musk Mellon.
Escaped to waste grounds. Mason: banks of the Ohio near Point Pleasant. Monongalia: waste grounds, near Morgantown.

MICRAMPELIS, Raf.

M. echinata, (Muhl.), Raf. (*Echinocystis lobata*, T & G.)
Escaped from gardens, where it is frequently grown as a vine for fences and rock work ornamentation.

SICYOS, L.

S. angulatus, L. Star Cucumber. M. & G.
Damp places. Monongalia: along Decker's Creek; near Easton? Fayette: near Nuttallburg—L. W. N.

CACTACEÆ.

OPUNTIA, Mill.

O. vulgaris, L. Prickly Pear.
Open fields and among scrub pines in the Devonian formations of Hardy: near Moorefield, where it is a prevalent weed in many places.

FICOIDEÆ.

MOLLUGO, L.

M. verticillata, L. Carpet-weed. M. & G.
Waste and cultivated grounds. Monongalia: hills about Morgantown. Marion: near Fairmont. Fayette: near Nuttallburg, sandy banks of New River—L. W. N.

UMBELLIFERÆ.

HYDROCOTYLE, L.

H. Americana, L. Water Pennywort. M. & G.
Along streams. Jefferson: near Flowing Spring. Randolph: above Cricard P. O, Grant: near Bayard. Greenbrier: near White Sulphur Springs.

ERYNGIUM, L.

E. aquaticum, L. Rattle-snake Master. (*E. yuccæfolium*, Michx.)
Swampy places. Webster: at Welsh Glade.

DAUCUS, L.

D. CAROTA, L. Wild Carrot. "Devil's Plague." G., M. & G.
Fields, meadows, and roadsides. Lewis, Randolph,
Monongalia. Marion, Berkeley, Morgan, Mineral, Preston,
Grant, and Kanawha counties. Gilmer: near Glenville—V.
M. Fayette: near Nuttallburg—L. W. N. Jefferson: near
Shenandoah Junction; near Charlestown; and Summit
Point. Greenbrier: near Caldwell and White Sulphur
Springs. Summers: near Greenbrier Stock Yards, and Hin-
ton. Mason: near Point Pleasant. Mercer: near Ingleside;
and reported from every county in the State.

Forma **rosea.**
With rose colored flowers, a frequent form in Monongalia: near Morgantown: and along the Kingwood Pike.

ANGELICA, L.

A. Curtisii, Buckley.
Sandy river banks. Monongalia: near Camp Eden.
Preston: near Reedsville. Grant: near Bayard. Greenbrier:
near White Sulphur Springs.

A. villosa (Walt.), B. S. P. Hairy Angelica. *A. hirsuta,* Muhl.
Frequent in dry woods and glady meadows. Web-
ster: Long and Welsh Glades. Jackson: near Sandyville.
Tucker: near Davis. Fayette: near Nuttallburg—L. W. N.
Monongalia: near Camp Eden. Preston: near Terra Alta.
Randolph: on Point Mountain, alt. 2540 ft. Hardy: near
Moorefield – G.

A. atropurpurea, L. Purplish Angelica. *Archangelica atropurpurea,* Hoffm. M. & G.
Low grounds and river banks. Grant: near Bayard.
Preston: near Terra Alta. Tucker: near Davis. Fayette:
near Kanawha Falls—James.

TIEDEMANNIA, DC.

T. rigida, (L.). Coult. & Rose. Cowbane.
Swampy spots. Randolph: along Shaver's Fork of
Cheat. Fayette: near Nuttallburg—L. W. N. Preston:
near Terra Alta.

HERACLEUM, L.

H. lanatum, Michx. Cow Parsnip.
Wet grounds. Lewis: along Leading Creek. Upshur:
near Lorenz. Randolph: along Tygart's Valley River.
Webster: Welsh Glade.

PASTINACA, L.

P. SATIVA, L. Wild Parsnip. M. & G.
Waste grounds and cultivated fields. Randolph:
Cheat Mountains, alt. 3350 ft. Jefferson: near Shenandoah
Jc. Greenbrier: near White Sulphur Springs. Mineral:
opposite Cumberland. Morgan: near Hancock. Wood: near
Kanawha Station. Mason: near Point Pleasant. Hardy:
near Morefield.— G.

THASPIUM, L.

T. aureum, (L.). Nutt. Meadow Parsnip. L. W. N., M. & G.
Thickets and meadows. Frequent throughout the
State.

Var. **cordatum.** (Walt.). B. S. P. *var. trifoliatum, Gray,* in pt.
M. & G.
With the species, but probably more frequent. Har-
dy: near Moorefield—G.

T. barbinode, Nutt.
Rich woods. Mercer: plentiful near Bluestone Jc.

LIGUSTICUM, L.

L. actæfolium, Michx.
Rich soil. Grant: near Bayard, plentiful along the
Blackwater Fork of Cheat River.

PIMPINELLA, L.

P. integerrima, (L.). Bth. & Hook.
Rocky hillsides. Lewis: along Stone Coal Creek.
Randolph: on Point Mountain.

DEERINGIA, Adans. (1763).
(Cryptotænia, DC. 1829)

D. Canadensis, L. Honewort. *Cryptotænia Canadensis (L.)*
DC. M. & G.
Shady rocks. Monongalia: Roundbottoms and Little
Falls. Marion: near Opekiska. Fayette: near Nuttallburg
—L. W. N.

ZIZIA, Koch.

Z. cordata, Koch.
River banks. Fayette: near Nuttallburg—L. W. N.

Z. aurea, Koch.
Damp places. Monongalia: The Flats, and along the Monongahela. Marion: near Opekiska. Fayette: near Nuttallburg—L. W. N.

Z Bebbii, C. & R. *Z. aurea,* var. *Bebbii,* C. & R.
Woodlands. Fayette: near Nuttallburg—L. W. N.
Of specimens gathered in Mason, near Pt. Pleasant, Prof. Coulter says: "Flowers too white, and altogether it does not quite fit, and is indeterminable on account of the immature fruit."

CICUTA, L.

C. maculata, L. Spotted Cow-bane. Beaver Poison.
Swampy spots, and wet meadows. Randolph: along Tygart's Valley River plentiful. Fayette: along Loup Creek—James; near Nuttallburg—L. W. N. Putnam: near Scott's Depot—James. Gilmer: Glenville—Prof. Brown; near DeKalb P. O. Morgan: near Cacapon. Monongalia: near Ice's Ferry. Mason: near Point Pleasant.

C. bulbifera, L.
Wet places. Mason: near Pt. Pleasant.

CHÆROPHYLLUM, L.

C. procumbens, (L.) Crantz.
Ohio: Elm Grove, near Wheeling—M. & G.

OSMORHIZA, Raf.

O. Claytoni (Michx.). B. S. P. *O. brevistylis,* D C. M. & G.
Rich woods. Wirt: above Elizabeth. Gilmer: near Glenville—V. M.; Prof. Brown. Monongalia: opposite Roundbottoms. Grant: near Bayard. Fayette: near Nuttallburg—L. W. N.

O. longistylis (Torr.). DC. M. & G.
Rich Woods. Monongalia: near Morgantown. Marion: near Fairmont. Tucker: near Davis. Wirt: above Elizabeth.

ERIGENIA, Nutt.

E. bulbosa, Nutt. Harbinger of Spring. M. & G.
Rich open woods. Monongalia: opposite Granville, plentiful. Fayette: near Nuttallburg—L. W. N.

SANICULA, L.

S. Marylandica, L. Black Snake-root. M. & G.
Rich woods. Monongalia: near Morgantown. Preston: near Terra Alta. Fayette: near Nuttallburg—L. W. N.

S. Canadensis, L. M. & G.
Rich soil. Monongalia: near Little Falls. Marion: near Opekiska. Fayette: near Nuttallburg—L. W. N.

ARALIACEÆ.

ARALIA, L.

A. spinosa, L. Angelica Tree. Hercules' Club. M. & G.
Rich mountain woods. Webster: Buffalo Bull Mt. alt. 2595 ft., plentiful. Preston: near Rowlesburg. Summers: along the Greenbrier River; near Hinton. Fayette: near Nuttallburg—L. W. N.; at Gauley Bridge, abundant. Monongalia: near Morgantown.

A. racemosa, L. Spikenard. M. & G.
Deep, cold woods, frequent in the Alleghanies. Randolph: Cheat Mountains, alt. 3350 ft.; Point Mountain, alt. 3560 ft. Hampshire: Ice Mountain. Tucker: near Falls of Blackwater. Gilmer: near Glenville—V. M. Greenbrier: near White Sulphur Springs. Fayette: near Nuttallburg—L. W. N. Summers: near Hinton.

A. nudicaulis, L. Wild Sarsaparilla. M. & G.
Rich woods, frequent. Monongalia: the Flats, Rich Woods, and along the Monongahela. Marion: near Opekiska. Randolph: on Point Mountain. Grant: near Bayard.

A. quinquefolia (L.), Dec. & Pl. Ginseng. "Sang." M. & G.
Rich, deep woods. Wirt: near Burning Springs. Jackson: near Ripley. Gilmer: near Glenville—V. M. Grant: near Bayard. Randolph: Rich, Cheat and Point Mountains. (One store at Crickard P. O. buys from this neighborhood $1,500 worth annually of the mountaineers.) Webster: Buffalo Bull range. Fayette: near Nuttallburg— L. W. N. Summers: near Hinton.

A hispida, Vent.
Rocky soils. Tucker: near Davis, along Blackwater Fork of Cheat.

CORNACEÆ.

CORNUS, L.

C. florida, L. Flowering Dogwood. M. & G.
Dry woods. Monongalia throughout, some quite large trees near Morgantown. Wood, Wirt and Calhoun counties. Gilmer: near Glenville—Prof. Brown;—V. M. Lewis and Upshur counties. Randolph: on Cheat Mountains, alt. 3600 ft. Marion: Webster: Fayette: near Nuttallburg—L. W. N. Hardy; near Moorefield—G. Mercer: near Bluefield.

C. circinata, L'Her. Round-leaved Dogwood.
Damp, cool woods. Rare. Upshur: near Lorentz.

C. sericea, L. Kinnikinnik.
Wet places. Grant: near Bayard. Randolph: along Tygart's Valley River. Nicholas: along Peter Creek. Fayette: near Nuttallburg—L. W. N.

C. candidissima, Marsh. Panicled Cornel. *C. paniculata*, L'Her.
Thickets and river banks. Monongalia: near Morgantown. Marion: Montana; along Beaver Creek. Randolph: Cheat River, alt. 2700 ft. Summers: near Hinton. Hampshire: near Romney.

C. alternifolia, L. f.
Hillside copses. Monongalia, Marion, Preston, Wood and Calhoun counties. Gilmer: near Glenville--V. M. Lewis: along Leading Creek. Upshur: near Lawrence. Fayette: near Nuttallburg—L. W. N. Greenbrier: near White Sulphur Springs. Summers: near Hinton.

NYSSA, L.

N. sylvatica, Marsh. Black Gum. M. & G.
Various situations throughout the State. Wood: near Leachtown. Wirt: along Straight Creek. Calhoun: near Brookville. Gilmer: near Glenville—V. M. Monongalia: near Morgantown. Randolph: on Point Mountain. Webster: on Buffalo Bull Mountains. Fayette: near Nuttallburg—L. W. N. Summers: near Hinton; and along the Greenbrier River. Kanawha: near Handley. Mercer: near Ingleside.

CAPRIFOLIACEÆ.

SAMBUCUS, L.

S. Canadensis, L. Common Elder. L. W. N., V. M., M. & G.
Rich soils; common in bottoms and along fences
throughout the State; even in the pine and spruce forests of
the higher mountains; altitude on Point Mountain 3050 ft.

S. racemosa, L. Red-berried Elder. M. & G.
Deep, rich mountain woods, near rivulets. Abund-
ant in Randolph, Grant and Tucker counties. Fayette: near
Nuttallburg—L. W. N.

Forma **albicocca,** Britt.
With the species rare. Randolph: on Point Mountain. Grant: near Bayard.

VIBURNUM, L.

V. lantanoides, Michx. Hobble-bush.
Cold, rich ravines. Randolph: near the summit of
Point Mountain. Grant: near Bayard. Tucker: along the
Blackwater.

V. acerifolium, L. Arrow Wood. Dockmackie. M. & G.
Cool, rocky woods. Throughout the móuntains of
the eastern counties. Preston and Monongalia: along Cheat
River. Gilmer: near Glenville—V. M. Mineral: near
Keyser—W. Fayette: near Nuttallburg—L. W. N. Grant:
near Bayard.

V. dentatum, L. Arrow-wood.
Wet places or damp thickets. Upshur: near the Sum-
mit on Staunton Pike. Fayette: near Nuttallburg, rare—
L. W. N.

V. nudum, L.
Rich woods. Randolph: at Ford's, near the Middle
Fork River. Webster: Upper Glade.

V. Lentago, L. Sweet Viburnum. Sheep-berry.
Rich banks of streams. Randolph: on Point Mountain. alt. 3660 ft.

V. prunifolium, L. Black Haw. Nanny-berry.
Copses and edges of woods. Wirt: near Burning
Springs. Mineral: near Keyser—W. Gilmer: near Glen-
ville—V. M.; Prof. Brown. Fayette: near Nuttallburg—L.
W. N. Summers: near Hinton.

TRIOSTEUM, L.

T. perfoliatum. L. Tinker's Weed. Wild Coffee.
Rich borders, infrequent. Randolph: Cheat Mts., alt.
4,600 feet. Gilmer: near Glenville—V. M. Fayette: near
Nuttallburg—L. W. N. Hardy: near Moorefield—G. Mercer: near Princeton.

SYMPHORICARPOS, Juss.

S. orbiculatus, Moench.   *S. vulgaris,* Michx.
Dry places. Nicholas: near Peter Creek. Fayette:
near Nuttallburg, alt. 2,000 ft.—L. W. N.

LONICERA, L.

L. glauca, Hill.   Smooth Honeysuckle.   *L. parviflora,* Lam.
Rocky soils. Monongalia: near Morgantown.

L. JAPONICA, Thunb.
Escaped from cultivation. Mason: banks of the Ohio
near Pt. Pleasant. Jefferson: near Shepherdstown. Taylor:
near Grafton.

DIERVILLA, Tourn.

D. trifida, Moench.   Bush Honeysuckle.
Thickets.   Monongalia: near Morgantown. along
Decker's Creek.

R U B I A C E Æ.

HOUSTONIA, L.

H. cærulea, L.   Bluets.   Innocents.   M. & G.
Moist fields. Monongalia: Marion: Preston: Wood;
Wirt: Calhoun: Lewis: and Upshur. Gilmer: near Glen-
ville—Prof. Brown—V. M. Kanawha—James. Mineral:
Jefferson: Berkeley: and Morgan. Fayette: near Nuttall-
burg—L. W. N. Hardy: near Moorefield—G.

Forma **albiflora.**
Grassy places, Permian formations. Monongalia:
near Cassville.

H. serpyllifolia, Michx.
Rocky places. Tucker: rocks below the falls of Blackwater. Monongalia: on rocks in Tibb's Run.

H. purpurea, L.
Wooded opens. Gilmer: near Glenville—V. M. Fayette: near Nuttallburg—L. W. N.; Kanawha Falls—James. Hampshire: near Romney. Greenbrier: near White Sulphur Springs. Summers: near Hinton. Monongalia: near Ice's Ferry.

Var. **ciliolata,** Gray.
Monongalia: near Morgantown. Fayette: near Nuttallburg—L. W. N. Greenbrier: near White Sulphur Springs.

Var. **longifolia** (Gaertn.). Gray. M. & G.
Dry soils, the most common form of the species. Wood: near Lockhart's Run. Wirt: near Burning Springs. Calhoun: near Grantsville. Gilmer: near DeKalb. Lewis: up Stone Coal Creek. Upshur: near Buckhannon. Randolph: near Cricard P. O. Cabell: near Barboursville—James. Fayette: near Nuttallburg—L. W. N. Mercer: near Bluefield.

Var. **tenuifolia,** Gray.
Greenbrier: near White Sulphur Springs.

Var. **calycosa,** Gray.
Greenbrier: near White Sulphur Springs.

CEPHALANTHUS, L.

C. occidentalis, L. Button-bush. M. & G.
Along streams. Monongalia: along the Monongahela and Cheat Rivers. Preston: general in the glades and along streams. Grant: near Bayard. Randolph: along Tygart's Valley River; near Cheat Bridge. Fayette: near Nuttallburg L. W. N.; near Kanawha Falls—James. Monroe—near Alderson. Summers: near Rifle and Hinton.

MITCHELLA, L.

M. repens, L. Partridge-berry. M. & G.
Rich woods, under evergreens. Upshur: Sand Creek. Grant: near Bayard. Tucker: along Blackwater. Mineral: Knobby Mts.—W. Randolph: along Cheat River. Gilmer: near Glenville—V. M. Fayette: near Nuttallburg—L. W. N. Kanawha: near Coalburg—James.

DIOIDA, L.

D. teres, Walt. Button-weed.
Sandy river banks. Ohio: along Bogg's Run, near Wheeling—M. & G. Preston: banks of Cheat River. Fayette: near Nuttallburg, rare—L. W. N.

GALIUM, L.

G. Aparine, L. Goose-grass. Cleavers. G.,L.W. N.,V.M.,M.& G.
Shaded places. Frequent throughout the State.

G. pilosum, Ait.
Dry copses. Fayette: near Nuttallburg—L. W. N.; near Kanawha Falls—James. Kanawha: near Coalburg—James. Monongalia: near Little Falls and Uffington; near Camp Eden.

G. circæzans, Michx. Wild Liquorice.
Rich woods. Wood: near Lockhart's Run. Monongalia: Rich Woods near Morgantown; Ice's Ferry and Camp Eden. Fayette: near Nuttallburg—L. W. N.

G. laceolatum, Torr.
Dry woods. Monongalia: near Morgantown. Ohio: near Wheeling—M. & G. Fayette: near Nuttallburg—L. W. N.

G. latifolium, Michx.
Fayette: near Nuttallburg, uncommon—L. W. N. Preston: near Rowlesburg—M. & G.

G. trifidum, L. Small Bedstraw.
Low, wet grounds. Monongalia, Lewis, Upshur, Gilmer, Calhoun, Wirt, Wood, and Webster: Long Glade. Fayette: near Nuttallburg—L. W. N.

Var. **latifolium,** Torr.
Damp soils. Webster: in Long Glade.

G. concinnum, Torr & Gray. M. & G.
Low, wet grounds. Wood: near Kanawha Station. Wirt: near Elizabeth. Lewis: along Leading Creek. Randolph: near Valley Bend.

G. asprellum, Michx. Rough Bedstraw.
Alluvial bottoms. Monongalia: along the Monongahela River. Greenbrier: near White Sulphur Springs—M. &G.

G. triflorum, Michx.   Sweet-scented Bedstraw.
Rich woodlands. Lewis: along Leading Creek. Upshur: near Lorentz. Webster: along Buffalo Bull range.
Monongalia: near Morgantown. Fayette: near Nuttallburg
—L. W. N.

. VALERIANACEÆ.

VALERIANA, L.

V. pauciflora, Michx.   Valerian.
Fields and open woods.   Ohio: near Moundsville—
M. & G.

DIPSACEÆ.

· DIPSACUS, L.

D. SYLVESTRIS, Mill.   Teasel.   "Water Thistle."   Huttonweed." M. & G.
Roadsides and waste places. Wirt: near Burning
Springs and Elizabeth.   Marion: near Worthington; near
Fairmont and Houghton in great quantity. Webster: Buffalo Bull Mountains, alt. 2100 ft. Fayette: near Crescent;
near Nuttallburg—L. W. N.   Kanawha: along the Kanawha
and Pocotaligo Rivers. Jackson: along Allen's Fork. Gilmer: near Glenville—V. M.   Jefferson: near Flowing Spring
and Shenandoah Jc. Randolph: Cheat Mts. near Cheat Bridge,
alt. 2700 ft.; near Huttonsville. Greenbrier: near White
Sulphur Springs, near Fort Spring. Monroe: near Alderson.
Summers: near Hinton. Monongalia: along Decker's Creek.
Harrison: near Lumberport. Mineral: opposite Cumberland. Berkeley: near Martinsburgh. Hardy: near Moorefield. Mercer: near Ingleside, and Ada.

COMPOSITÆ.

ELEPHANTOPUS, L.

E. Carolinianus, Willd.
Dry banks. Fayette: near Nuttallburg—L. W. N.

E. tomentosus, L.   "Tobacco Weed."   "Devil's Grandmother."
Fields. Harrison: near Quiet Dell. Upshur: near
Lorentz.

VERNONIA, Schreb.

V. altissima, Nutt. Iron-weed
Low grounds. A frequent weed throughout the northern, central, and western portions of the State. Fayette: near Nuttallburgh—L. W. N.

V. Noveboracensis (L.), Willd. Iron Weed. G., L. W. N., M. & G.
In meadows and pastures, common throughout the State.

Var. **latifolia,** Gray.
Meadows and fields. Mason: near Point Pleasant. Monongalia: near Morgantown. Fayette: near Nuttallburg —L. W. N.

EUPATORIUM. L.

E. purpureum, L. Queen of the Meadow. "Quill-wort." L. W. N. V. M., M. & G., G.
Low grounds. Common throughout the State. Cheat Mountains in Randolph at an altitude of 3600 feet.

Var. **amœnum,** Gray.
Rich woods along runs. Grant: Buffalo Creek near Bayard· Tucker: Beaver Creek near Davis.

E. hyssopifolium, L.
Sterile soil. Jefferson: near Shepherdstown.

E. rotundifolium, L., *var.* **pubescens.** (Muhl.), B. S. P.
(*E. pubescens,* Muhl.)

Dry hillsides. Fayette: near Nuttallburg—L. W. N. Jefferson: near Shepherdstown. Monongalia: near Morgantown and Camp Eden.

E altissimum, L. Tall Boneset.
Dry soils. Monongalia: near Little Falls and Beechwoods.

E. sessilifolium, L. Upland Boneset.
River banks. Monongalia: near Beechwoods. Fayette; near Nuttallburg, plentiful—L. W. N.

E. perfoliatum, L. Boneset. Thorough-wort. G., L. W. N., V. M., M. & G.
Damp places. Common throughout the State.

E. ageratoides, L. White Snake-root.
Rich woods. Monongalia: along Decker's Creek and elsewhere plentiful. Randolph: Cheat Mountains near Cheat Bridge. Marion: near Worthington. Fayette: near Nuttallburg--L. W. N. Hardy: near Moorefield—G.

E. aromaticum, L.
Rich soil. Fayette: near Nuttallburg. Hardy: near Moorefield--G.

E. cœlestinum, L. Mist-flower. M. & G.
Rich soils. Putnam, Jackson, Wood and Monongalia. A common weed. Randolph: along Tygart's Valley River. Harrison: along the "Monongah" R. R. Summers: near Hinton. Fayette: near Nuttallburg--L. W. N. Putnam: near Buffalo. Kanawha: near Charleston. Mason: near Point Pleasant. Marion: near Montana and Worthington. Jefferson: near Shepherdstown.

LACINARIA, Hill (1762).
Liatris, Schreb. (1791).

L. spicata, (L.), OK.
Among rocks, banks of New River—Selby. Fayette: near Nuttallburg, heads 5-flowered--L. W. N.

CHRYSOPSIS, Nutt.

C. Mariana (L.), Nutt.
Dry, rocky roadside. Fayette:R. & K.turnpike near Nuttallburg—L. W. N.

SOLIDAGO, L.

S. latifolia, L. M. & G.
Moist, shaded banks. Monongalia: banks of the Monongahela and Cheat River. Fayette: near Nuttallburg—L. W. N.

S. cæsia, L. L. W. N., V. M., M. & G.
Rich woodlands. Frequent throughout the State.

S. Curtisii, Torr. & Gray.
Woodlands. Fayette: near Nuttallburg, common— L. W. N.

S. bicolor, L. L. W. N., M. & G.
Dry fields and copses. Frequent throughout the State.

S. monticola, Torr. & Gray.
Woods and opens. Fayette: near Nuttallburg, alt.
2000 ft—L. W. N.

S. puberula, Nutt.
Sunny opens. Fayette: near Nuttallburg—L. W. N.

S. speciosa, Nutt.
Cliffs and banks. Fayette: near Nuttallburg—L.
W. N.

S. odora, Ait Sweet Golden-rod.
Fayette: near Nuttallburg—L. W. N.

S. rugosa, Mill.
Borders of fields and copses. Along Cheat River.
Randolph, Tucker, Preston and Monongalia counties. Fayette: near Nuttallburg—L. W. N. Shores of the Monongahela in Barbour, Taylor and Marion counties.

S. ulmifolia, Muhl.
River banks. Ohio: Thomas Hill near Wheeling—M.
& G. Brooke: M. & G. Fayette: near Nuttallburg—L.
W. N.

S. Boottii, Hook.
Dry open woods. Putnam: near Buffalo. Fayette:
near Nuttallburg—L. W. N.

S. arguta, Ait.
River banks. Ohio: banks of the Ohio River near
Wheeling—M. & G.

S. juncea, Ait. "Yellow Top."
Fields and waste places. Common throughout the
northern, central and western counties. Fayette: near Nuttallburg—L. W. N. Berkeley: near Martinsburg. Mason:
near Point Pleasant. Hardy: near Moorefield—G.

Var. **scabrella,** Gray.
With the species. Frequent.

Var. **ramosa,** Britten.
River banks. Monongalia: near Morgantown, below
highwater mark along the Monongahela.

S. serotina, Ait.
Fayette: near Nuttallburg—L. W. N. Monongalia:
near Morgantown.

var. **gingantea,** (Ait.), Gray.
Thickets. Gilmer: near Glenville—V. M. Preston: near Rowlesburg.

S. rupestris, Raf.
Rocky river banks. Fayette: along the Gauley at Gauley Mountain; Kanawha Falls and Hawk's Nest—James.

S. Canadensis, L. L. W. N., M. & G.
Borders and waste fields. Common throughout the State.

S. nemoralis, Ait.
Dry, sterile fields. Fayette: near Nuttallburg—L. W. N. Common throughout the northern counties.

S. lanceolata, L.
River banks. Along Cheat River throughout its length. Along the Monongahela in Marion, Taylor and Monongalia counties. Gilmer: along the Little Kanawha— V. M. Mason: near Point Pleasant.

S. Caroliniana, (L.), B. S. P. *S. tenuifolia,* Pursh.
Sandy fields. Monongalia: near Morgantown.

SERICOCARPUS, Nees.

S. asteroides (L.), B. S. P. White-topped Aster. M. & G.
Dry grounds. Frequent or common throughout the State. Kanawha: near Charleston—James. Greenbrier: near White Sulphur Springs. Fayette: near Nuttallburg— L. W. N.

BRACHYCHÆTA, T. & G.

B. cordata, Torr. & Gray.
Dry woods. Fayette: near Nuttallburg, plentiful— L. W. N.

ASTER, L.
A. corymbosus, Ait. M. & G.
Shady places. Monongalia: near Morgantown. Fayette: near Nuttallburg—L. W. N.

A. macrophyllus, L.
Open woods. Fayette: near Nuttallburg—L. W. N.

A. patens, Ait.
Rocky river banks. Fayette: near Nuttallburg—L. W. N. Summers: near Hinton.

Var. **phlogifolius,** (Muhl.), Nees.
> Open woods. Fayette: near Nuttallburg--L. W. X.

A. lævis. L.
> Rocky river banks. Monongalia: near Little Falls.
> Fayette: near Nuttallburg, plentiful--L. W. X.

A. undulatus. L.
> Dry woods. Mason: near Point Pleasant. Kanawha:
> near Charleston. Fayette: near Nuttallburg, common-- L.
> W. X.

A. cordifolius, L.
> Woodlands. Monongalia: near Morgantown and Lit-
> tle Falls. Fayette: near Kanawha Falls—James; near Nut-
> tallburg L. W. X. Mason: near Point Pleasant.

Var. **lævigatus,** Porter.
> Woodlands and opens. Monongalia: near Morgan-
> town, abundant.

A. virgatus, Ell.
> Rocky river banks. Fayette: near Nuttallburg-- L.
> W. X. Preston: along Cheat River. Monongalia: near
> Camp Eden.

A. ericoides. L.
> Dry open places. Fayette: near Nuttallburg –L. W.
> X. Mason: near Point Pleasant. Wood: near Parkersburg.
> Monongalia: near Morgantown.

Var. **pusillus,** Gray.
> Dry fields. Monongalia: plentiful about Morgan-
> town.

Var. **villosus,** Torr. & Gray
> Roadsides, etc. Fayette: near Nuttallburg, common
> —L. W. X. Monongalia: near Morgantown. Marion: near
> Fairmont.

A. lateriflorus, (L.). Britt.   *A. miser,* Man.   *A. diffusus,* Ait.
> Dry or moist grounds : Monongalia: near Morgantown.
> Frequent throughout the northern counties. Hardy : near
> Moorefield—G.

Var. **hirsuticaulis,** (Lind.).   "Nail-rod."
> Fields and roadsides. Cabell: near Barbour-ville.
> Monongalia: near Morgantown: and common throughout
> the northern, central and western counties.

A. multiflorus. Ait.
Hardy: near Moorefield—G.

A. dumosus. L.
Hardy: near Moorefield—G.

A. vimineus, Lam.
Shaded roadsides and fields. Fayette: near Nuttallburg, altitude 2000 ft., plentiful—L. W. N. Monongalia:
near Morgantown. Mason: near Point Pleasant.

Var. **foliolosus,** Gray.
Monongalia: near Morgantown, Uffington and Little
Falls, common.

A. paniculatus. Lam.   (*A. simplex,* Willd.)
Low grounds. Fayette: near Nuttallburg—L. W. N.
Mason: near Point Pleasant. Putnam: near Buffalo.

A. salicifolius, Ait.
Near streams. Monongalia and Preston: banks of
Cheat River.

A. Novi-Belgii, L.
Damp meadows. Monongalia: near Morgantown.

A. prenanthoides, Muhl.
Rich woods and borders of streams. Randolph: Cheat
Bridge, alt. 3360 ft. Monongalia: shore of Monongahela
above Morgantown. Fayette: near Nuttallburg L. W. N.

A. puniceus, L.
Swampy places. Fayette: near Nuttallburg, uncommon—L. W. N. Hardy: near Moorefield—G.

A. umbellatus, Mill.
Moist thickets. Along Cheat River in Randolph,
Tucker, Preston and Monongalia counties. Fayette: near
Nuttallburg—L. W. N.

A infirmus, Michx.
Mountain woods. Randolph: Point Mountain, alt.
2800 ft. Fayette: near Nuttallburg—L. W. N.

A. acuminatus, Michx.
Cool, rich woods. Randolph: near Cheat Bridge.
Fayette: near Kanawha Falls—James.

[**A. tenuifolius,** L.   (*A. flexuosus,* Nutt.)   M. & G.]

A. linariifolius, L.
Rocky places. Fayette: near Nuttallburg, along the banks of New River below high water mark, common—L. W. N.

ERIGERON, L.

E. Canadensis, L. Butter-weed. Horse-weed. L. W. N., V. M.,
M. & G.
Waste places. Common throughout the State.

E. annuus, (L.). Pers. Daisy Fleabane. Sweet Scabious. L. W. N.
· V. M., M. & G.
A weed in meadows and fields. Common through out the State.

E. ramosus (Walt.), B. S. P. Daisy Fleabane. (*E. strigosus*, Muhl.)
Fields and waste places. Monongalia: the Flats and Uffington. Fayette: Nuttallburg L. W. N.

E. pulchellus, Michx. Robin's Plantain. (*E. bellidifolius*, Muhl.)
L. W. N., M. & G.
Copses, common throughout the State.

E. Philadelphicus, L. Common Fleabane.
Moist ground. Frequent throughout the northern counties.

ANTENNARIA, Gærtn.

A. plantaginifolia (L.), Hook. Everlasting. L. W. N., M. & G.
Sterile hills. Frequent or common throughout the State.

ANAPHALIS. DC.

A. margaritacea (L.), Bth. & Hook. Pearly Everlasting. M. & G.
Dry hills and woods. Monongalia: along Decker's Creek. Marion: above Opekiska.

GNAPHALIUM, L.

G. obtusifolium, L. Everlasting. (*G. polycephalum*, Michx.
M. & G.
Old fields. Frequent or common throughout the northern and central counties. Fayette: near Nuttallburg—L. W. N. Hardy: near Moorefield—G.

G. ulignosum, L. Low Cud-weed. M. & G.
Low grounds. Grant: near Davis. Gilmer: near
Glenville—V. M.. Prof. Brown. Monongalia: near Morgantown. Mason: near Point Pleasant. Wood: near Parkersburg.

G. purpureum, L. Purplish Cud-weed.
Sandy soil. Monongalia: near Beechwoods and Ice's
Ferry. Fayette: near Nuttallburg—L. W. N.

INULA, L

I **HELENIUM,** L. Elecampane. M. & G.
Fields. Wirt: near Burning Springs. Upshur: near
Lorentz. Nicholas: along Mumble-the-peg Creek. Fayette:
near Nuttallburg—L. W. N. Greenbrier: near Ronceverte.
Jefferson: near Shepherdstown. Hampshire: near Romney.
Monongalia: near Stumptown.

POLYMNIA, L.

P. Canadensis. L. Leaf Cup. M. & G.
Moist shaded ravines. Fayette: near Kanawha Falls
and Hawk's Nest—James; Porter: near Nuttallburg – L. W.
N. Hardy: near Moorefield—G.

Var. **radiata,** Gray.
Rich rocky soil. Fayette: near Nuttallburg—L.W.N.

P. Uvedalia, L.
Rich soil. Randolph: frequent along Tygart's Valley River. Fayette: near Nuttallburg—L. W. N.

SILPHIUM, L.

S. Asteriscus, L.
Dry sandy soil. Wirt: beyond Burning Springs.
Jackson: near Ripley.

S. trifoliatum, L. Rosin-weed.
Dry hills and banks. Fayette: near Nuttallburg—L.W.N.

S. perfoliatum, L. Cup Plant.
Along streams. Fayette: near Hawk's Nest—James;
near Nuttallburg—L. W. N.

CHRYSOGONUM, L.

C. Virginianum. L.
Dry soils. Hardy: near Moorefield—G.

PARTHENIUM, L.

P. integrifolium, L.   Sneeze-wort.
Dry soils.   Fayette: near Nuttallburg, banks of New
River below high water mark, plentiful—L. W. N.   Green-
brier: near White Sulphur Springs—M. & G.

AMBROSIA, L.

A. trifida, L.   Great Rag-weed.   G., L. W. N., M. & G.
Moist places.   Common or abundant throughout the
State.

Var. **integrifolia,** (Muhl.), T. & G.
With the species, uncommon.   Monongalia: near
Morgantown.   Wood: near Parkersburg.   Fayette: near
Nuttallburg—L. W. N.   Berkeley: near Martinsburg.

A. artemisiæfolia, L.   Rag-weed.   G., L. W. N., V. M., M. & G.
Fields and roadsides.   Abundant throughout the
State.

XANTHIUM, L.

X SPINOSUM. L.   Spiny Clotbur.
Waste lands along rivers.   Kanawha: at Stockton's.
Mineral: near Piedmont.   Jefferson: near Shepherdstown.
Wood: near Parkersburg.   Berkeley: near Martinsburg.

X. STRUMARIUM, L.   Clotbur.   Cockle-bur.
Low waste grounds.   Monongalia, Marion and Gilmer
counties.   Wood: near Parkersburg.   Lewis: near Weston.
Jefferson: near Shepherdstown.

X. CANADENSE. Mill.   L. W. N.
Low waste grounds.   Common throughout the State.

ECLIPTA, L.

E. alba. (L.). Hassk.   (*Eclipta procumbens*, and *E. erecta*, Michx.)
Wet river banks.   Mason: banks of the Ohio near
Point Pleasant.   Ohio: near Wheeling   M. & G.   Fayette:
R. R. bank, Nuttallburg—L. W. N.

HELIOPSIS, Pers.

H. scabra. Dunal.   Ox-eye.
Fields.   Gilmer: near Glenville—V. M.

H. lævis, Pers.
Fayette: near Nuttallburg—L. W. N.

ECHINACEA, Moench.

E. PURPUREA, Moench. Purple Cone-flower.
Along the C. & O. R. R. Fayette: near Nuttallburg;
a rough, bristly form—L. W. N. Adventive from the west.

RUDBECKIA, L.

R. laciniata, L. Cone-flower. M. & G.
Low grounds. Monongalia: Little Falls, Beechwoods,
Uffington. and Morgantown. Fayette: near Nuttallburg—
L. W. N.

Var. **humilis,** Gray.
Monongalia: banks of Monongahela River below Morgantown.

R. fulgida, Ait.
Fields and Meadows. Monroe: abundant near Alderson. Hardy: near Moorefield G.

R. triloba. L. Brown-eyed Susan. M. & G.
Dry fields. Gilmer: near Glenville—V. M. Greenbrier: near White Sulphur Springs.

R. HIRTA, L. "Nigger Head." "Yellow Daisy." Brown-eyed Susan.
M. & G.
Becoming too frequent in Meadows. Randolph: Cric-
card P. O. Throughout the Ohio River counties. Fayette:
along Loup Creek—James, 1887: near Nuttallburg—L. W.
N. Wood: near Kanawha Station.

R. speciosa, Wender.
Dry soils. Ohio: near Wheeling—M. & G.

HELIANTHUS, L.

H. lœtiflorus, Pers.
Dry Opens. Fayette: near Nuttallburg.

H. occidentalis, Riddell. Western Sunflower.
Banks of New River. Fayette near Nuttallburg. infrequent—L. W. N.

Var. **Dowellianus,** T. & G.
Dry soils. Fayette: near Nuttallburg—L. W. N.

H. tomentosus. Michx.
Banks of New River. Fayette: near Nuttallburg—L. W. N.

H. grosse-serratus, Martens. Large-toothed Sunflower.
Dry fields. Upshur: near Buckhannon.

H. giganteus, L. Giant Wild Sunflower.
Low grounds. Randolph: near Cheat Bridge. Fayette: near Nuttallburg, plentiful—L. W. N. Preston: near Terra Alta.

H. lævigatus, Torr. & Gray.
Thickets. Preston: near Terra Alta.

H. doronicoides, Lam.
Dry grounds. Ohio: on Bogg's Island.—M. & G. Hardy: near Moorefield—G.

H. parviflorus, Bernh.
Thickets. Summers: near Hinton. Greenbrier: near White Sulphur Springs. Fayette: near Nuttallburg—L. W. N. Preston: near Terra Alta.

H. divaricatus, L.
Thickets and dry places. Fayette: near Nuttallburg —L. W. N. Jackson: up 8 mile creek.

H. hirsutus, Raf.
Dry banks. Fayette: near Nuttallburg, rare—L. W. N. Mason: Banks of the Ohio near Point Pleasant. Hardy: near Moorefield—G.

H. strumosus, L.
River banks and low copses. Monongalia: along Decker's Creek.

H. tracheliifolius, Willd.
Mountain Woods. Fayette: near Nuttallburg, uncommon—L. W. N.

H. decapetalus, L.
Rich open woods. Monongalia: near Little Falls and Uffington. Fayette: near Nuttallburg, the most common species here; petals mostly 8—L. W. N.

VERBESINA, L.

V. occidentalis, Walt.   Crownbeard.
Rich soil. Fayette: near Nuttallburg—L. W. N.; and
along the Great Kanawha River to its mouth. Jackson: up
8-Mile Creek. Wood: near Lockhart's Run. Monongalia:
near Morgantown. Summers: near Hinton. Jefferson:
near Shepherdstown. Berkeley: near Martinsburg.

RIDANIA, Adans. (1763)
(Actinomeris, Nutt. 1818)

R. alternifolia, (L.). OK.   (*Actinomeris squarrosa*, Nutt.)
Rich soil. Ohio: near Wheeling—M. & G. Fayette:
near Nuttallburg, common—L. W. N. Monongalia: near
Morgantown. Kanawha: near Charleston.

COREOPSIS, L.

C. lanceolata, L., *Var.* **villosa**, Michx.
Rich soil. Fayette: banks of New River near Nuttallburg—L. W. N.

C. pubescens, Ell
Rich shady place. Fayette: near Nuttallburg—L.
W. N.

C. trichosperma, Michx.
Fields. Kanawha: near Charleston. Monongalia:
near Morgantown.

C. auriculata, L.
Rich banks. Fayette: near Nuttallburg—L. W. N.;
near Hawk's Nest—Porter. Monroe: near Alderson.

C. senifolia, Michx.
Shady woods. Greenbrier: near White Sulphur
Springs—M. & G.

Var. **stellata**, Torr. & Gray.
Fayette: banks of New River near Nuttallburg—L.
W. N.

C. tripteris, L.   Tall Coreopsis.
Rich ground. Jackson: plentiful along 8-Mile Creek
and on Limestone Ridge. Fayette: near Nuttallburg—L.
W. N. Monongalia: near Little Falls.

BIDENS, L.

B. frondosa, L. Beggar's Ticks. Stick-tight. "Pitch-forks."
L. W. N., M. & G.
Damp waste places. Common throughout the State.

B. connata, Muhl. Swamp Beggar's Tick. M. & G.
Wet places. Frequent throughout the State.

Var. **comosa,** Gray.
Damp open places. Fayette: near Nuttallburg—L.
W. N. Monongalia: near Morgantown and frequent through-
out the State.

B. lævis (L.). B. S. P. *B. chrysanthemoides*, Michx. G., L. W. N.,
M. & G.
Wet places. Frequent throughout the State.

B. bipinnata, L. Spanish Needles. L. W. N., M. & G.
Dry places. Abundant throughout the State.

GALINSOGA, Ruiz & Pav.

G. PARVIFLORA, Cav.
Waste grounds. Mason: near Point Pleasant. Wood:
near Parkersburg.

HELENIUM, L.

H. autumnale, L. Sneeze-weed.
Alluvial river banks. Wirt: along the Little Kanawha River. Fayette: near Nuttallburg—L. W. N. Monongalia: near Morgantown. Randolph: near Cheat Bridge.
alt. 3660 ft. Summers: near Hinton. Hardy: near Moorefield—G.

ANTHEMIS, L.

A COTULA, L. Dogs Fennel. May-weed. L. W. N., M. & G.
Fields and waste grounds. Common throughout the
State.

A. ARVENSIS, L. Chamomile.
Waste places. Morgan: along the B. & O. R. R. near
No. 12 Water Tank.

ACHILLEA, L.

A. Millefolium, L. Yarrow. Milfoil. G., L. W. N., M. & G.
Common throughout the State, even in the most inaccessible portions of the virgin forests in the Alleghanies, where it certainly appears native. Randolph: Point Mountain, alt. 3300 ft. Nicholas: Buffalo Range, alt. 2875 ft.

CHRYSANTHEMUM, L.

C. Leucanthemum, L. Ox-Daisy. "Sheriff Pink." M. &. G.
Becoming too plentiful as a weed in fields, in the following counties: Monongalia, Marion, Hampshire: where it is often known as Sheriff Pink: Jackson, Preston, Kanawha: near Charleston—James; Cabell: near Barboursville—James (1877); Grant, Lewis, Upshur, Randoloph, Berkeley: near Martinsburg: Fayette: near Nuttallburg—L. W. N. Greenbrier: near Ronceverte, Caldwell, Fort Spring, and White Sulphur Springs. Hardy: near Moorefield—G. Mercer: near Princeton and Ingleside.

MATRICARIA, L.

M. DISCOIDEA, DC. Wild Chamomile.
Established on B. & O., R. R. bank, Morgan: near No. 12 Water Tank.

TANACETUM, L.

T. VULGARE, L. Tansy. M. & G.
Escaped to roadsides. Gilmer: near DeKalb. Lewis: near Weston. Grant: near Davis. Wood: near Parkersburg. Jefferson: near Shepherdstown. Monongalia: on Kingwood Pike.

SENECIO, L.

S. VULGARIS, L. Groundsel.
Roadsides, fence rows, streets, and waste places: adventive from Europe. Frequent.

S. aureus, L. Golden Rag-wort. L. W. N., V. M., M. & G.
Damp places in open woods. Frequent throughout the State.

Var. **obovatus** (Muhl.), T. & G.
Damp places. Lewis: near Weston. Monongalia: near Morgantown.

Var. **Balsamitæ** (Muhl.). T. & G.
Rocky open woods. Fayette: near Nuttallburg—L.
W. N. Monongalia: near Morgantown. Mercer: near
Beaver Spr.

CACALIA, L

C. suaveolens, L. Indian Plantain.
Rich banks. Monongalia and Marion: from Opekiska to Morgantown along the Monongahela River, frequent.
Preston: near Terra Alta. Summers: near Hinton Ohio:
near Wheeling—M. & G.

C. reniformis, Muhl. Great Indian Plantain.
Rich woods. Marion: along the F. M. & P., R. R.,
especially near Opekiska. Summers: near Greenbrier Stock
Yards. Monroe: near Alderson and Wolf Creek. Preston:
near Terra Alta. Ohio: Bogg's Island., near Wheeling—M.
& G.

C. atriplicifolia. L. Pale Indian Plantain.
Rich woodlands. Upshur: near Lorentz. Monon-
galia: banks of Cheat of Cheat River, near Camp Eden.
Ohio: near Wheeling—M. & G. Fayette: near Nuttallburg
—L. W. N.

ERECHTITES, Raf.

E. hieracifolia (L.), Raf. Fireweed.
Moist woods and banks, especially new fallows. Randolph: near Cheat Bridge, alt. 3700 ft. Fayette: near Nuttallburg—L. W. N. Monongalia: near Uffington and Morgantown.

ARCTIUM, L.

A. LAPPA, L. Burdock. L. W. N., M. & G.
Waste grounds, near dwellings. Abundant everywhere.

Var. **MINUS**. Gray.
Fayette: near Nuttallburg—L. W. N.

CNICUS, L.

C. LANCEOLATUS (L.), Willd. Common Thistle. L. W. N., M. & G.
Fields, waste grounds, and roadsides. Common.

C. altissimus, (L.). Willd.   Tall Thistle.

Fields and moist copses, frequent. Monongalia, Marion and Preston counties. Fayette: near Nuttallburg—L. W. N. Summers: near Greenbrier Stock Yards. Hardy: near Moorefield.

Var. **discolor.** Gray.   M. & G.

Fields. Jefferson: near Charlestown; Summit Point; and near Shepherdstown.

C. Virginianus, (Michx.). Pursh.   Virginia Thistle.

Woods and opens. Summers: near Hinton. Preston: near Terra Alta. Frequent throughout the State.

C. muticus. (Michx.). Pursh.   Swamp Thistle.   M. & G.

Wet places. Randolph: near Cheat Bridge, alt. 3700 ft. Upshur: near Lorentz. Kanawha: near Charleston. Preston: near Terra Alta.

C. odoratus, (Muhl.). B. S. P.   Pasture Thistle. *Cirsium pumilum.* Spr.

Dry fields. Greenbrier: near White Sulphur Springs. Preston: near Terra Alta; near Cranberry Summit—M. & G.

C. ARVENSIS (L.). Hoffm.   Canada Thistle.   M. & G.

Dry fields, becoming troublesome in many localities. Jefferson: plentiful near Charlestown, where it was doubtless brought in baled hay by the Federal troops during the war. Randolph: on the apex of Point Mountain, alt. 3700 ft., in a field owned and cultivated two years ago by a Connecticut gentleman, who probably brought the seed there from the east. Greenbrier: near White Sulphur Springs. Jefferson: near Summit Point; and Shenandoah Junction. Hancock: near Holliday's Cove. Brooke: at Wellsburg.

Reported also from: Hampshire: near Slanesville; and Capon Bridge. Brooke: near Wellsburg. Ohio: near Beech Glen School House. Summers: near Jumping Branch. Putnam: near Hurricane, Paradise and Confidence. Jefferson: near Summit Point, Middleway, Mohler's, Shenandoah Junction, Leetown, and Charlestown. Lewis: near Camden. Harrison: near Shinnston, and Wallace. Mineral: near Patterson's Depot (since destroyed). Berkeley: near Martinsburg, and Gerrardstown. Wirt: near Burning Springs. Wetzel: near Endicott. Jackson: near Sandy, and Silverton. Kanawha: near Pocotaligo, and Gazil. Mercer: near Concord Church. Wayne: near Stone Coal. Braxton: near Bulltown, and Tate Creek. Tyler: in Mead dist. Roane: near Newton, and Looneyville. Upshur: near Evergreen. Wood: near Murphy's Mills, Volcano, Parkersburg, and Rockport. Ritchie: near Berea. Fayette: near Mountain

Cove. Marshall: near Meighen. Hardy: near Wardensville. Preston: near Independence, 1889-91. Monroe: near Union. Greenbrier: near Trout Valley, and Lewisburg. Grant: near Greenland. Hancock: near Holliday's Cove. Taylor: near Grafton. Cabell: near Milton. Clay: near Valley Fork. Doddridge: near Leopold.

The presence of this weed in the localities noted in the second paragraph, where not corroborated in the first, is open to doubt.

ADOPOGON, Neck. (1790).

(Krigia, Schreb. 1791.)

A. Dandelion (DC.). Dwarf Dandelion.
Kanawha: near Charleston. (?)—James.

A. amplexicaulis. (Michx.)
Moist woods and opens. Monongalia: near Morgantown. Wood: near Lockhart's Run, becoming a bad weed—Hopkins.

CICHORIUM, L.

C. INTYBUS, L. Chicory.
Fields. Jefferson: two stations near Shepherdstown. Greenbrier: near White Sulphur Springs—M. & G.

HIERACIUM, L.

H. Canadense. Michx. Hawkweed.
Dry Woods. Webster: near Upper Glade.

H. paniculatum, L.
Moist grounds. Preston: near Cranberry Summit—M. & G. Fayette: near Nuttallburg—L. W. N.

H. venosum. L. Rattlesnake-weed. G., L. W. N., M. & G.
Openings, and edges of dry woods. Frequent throughout the State.

H. scabrum. Michx.
Dry open woods. Fayette: near Nuttallburg—L. W. N. Monongalia: along Decker's Creek. Preston: near Terra Alta.

H. Gronovii. L.
Dry soils. Fayette: near Nuttallburg, alt. 2000 ft. —L. W. N. Upshur, summit on Staunton Pike.

H. longipilum, Torr.
    Dry situations. Monongalia: Decker's Creek, near
Morgantown. Fayette: near Kanawha Falls, and Hawk's
Nest—James.

PRENATHES, L.

P. altissima, L.
    Rich moist woods.   Fayette: near Nuttallburg—L.
W. N.

P. alba, L.
    Open woods.   Hardy: near Moorefield—G.

P. serpentaria, Pursh.   Gall-of-the-Earth.
    Sandy woods.   Randolph: near Cheat Bridge, alt.
3550 ft.   Summers: near Hinton.   Marion: near Catawba.

TARAXACUM, Haller.

T. OFFICINALE. Web. Dandelion. (*T. Dens-leonis*, Desf.) L. W. N., G.
    All situations. Frequent throughout the State.

CHONDRILLA, Tourn.

C. JUNCEA, L.   "Naked-weed." "Skeleton-weed."
    Fields and roadsides. Hampshire: near Bloomery,
where the name Naked-weed has been given it on account of
the minutness of the leaves. Jefferson: near Summit, where
it is called Skeleton-weed, for the same reason; near
Charlestown.   Berkeley: near Martinsburgh.

LACTUCA L.

L. SCARIOLA, L.   Prickly Lettuce.
    Fields. Monongalia: near Laurel Point, where it has
become a troublesome weed.

L Canadensis, L.   Wild Lettuce. Horse-weed. "Devil-weed."
                                        L. W. N., M. & G.
    Meadows and fence-rows. Common throughout the
State.

L. integrifolia, Bigel.   "Devil's Iron-weed."
    Fields and roadsides. Monongalia: near Morgantown.
Mason: near Point Pleasant. Fayette: near Nuttallburg—
L. W. N. Jackson: near Douglas.

L. hirsuta,. Muhl.
>> Dry open mountain sides. Fayette: near Nuttall-
burg—L. W. N.

L. leucophæa, (Willd.), Gray.   (*Mulgedium leucophæum*, DC.)
>> Low woodlands. Fayette: near Nuttallburg—L. W.
N. Monroe: near Alderson.

L. villosa, Jacq.   (*Mulgedium acuminatum*, DC.)
>> Borders. Fayette: near Nuttallburg—L. W. N. Mon-
roe: near Alderson. Preston: near Terra Alta.

L. Floridana, (L.). Gaertn.
>> Open banks and borders of woods. Fayette: near
Nuttallburg—L. W. N.

SONCHUS, L.

S. OLERACEUS, L.   Sow-thistle.
>> Waste grounds. Ohio: near Wheeling—M. & G.

S ASPER, Vill.   Spiny leaved Sow-thistle.   M. & G.
>> Roadsides and wastes. Monongalia: near Morgan-
town. Narion: near Fairmont. Hampshire: near Slanesville. Wetzel: near Littleton. Lewis: near Vadis. Cabell:
near Union Ridge. Mercer: near Concord Church. Fayette:
near Nuttallburg—L. W. N. Doddridge: near Smithton.

TRAGOPOGON, L.

T. PORRIFOLIUS. L.   Salsify. Oyster-plant.
>> Waste grounds. Morgan: near No. 12 Water Tank.

CAMPANULACEÆ.

LOBELIA. L

L. cardinalis. L.   Cardinal Flower.
>> Low grounds, and low banks of streams. Nicholas:
Collett's Glade. Gilmer: near Glenville—V. M., Prof. Brown.
Randolph: near Cricard P. O. Greenbrier: near White Sulphur Springs. Summers: near Talcott, and Hinton. Kan
awha: near Kanawha City. Mason: near Brighton. Frequent throughout the State. Hardy: near Moorefield—G.

L. syphilitica, L.   Great Blue Lobelia.   M. & G.
>> Low wet grounds. Randolph: near Elkins, and along
the valley of Tygart's. Gilmer: near Glenville—V. M.
Fayette: near Nuttallburg. Greenbrier: near White Sulphur Springs. Monongalia: near Morgantown. Summers:

near Hinton. Jefferson: near Shepherdstown. Hardy: near Moorefield—G.

forma **albiflora,** Britt.
With the species. Randolph: near Huttonsville, frequent.

L. puberula, Michx.
Low grounds. Fayette: near Nuttallburg—L. W. N. Monongalia: near Morgantown.

L. amœna, Michx. *var.* **glandulifera.** Gray.
Swampy spots. Fayette: near Nuttallburg. alt. 2000 ft., rare—L. W. N.

L. leptostachys, A. DC.
Sandy soil. Wood: near Leachtown. Summers: near Hinton.

L. spicata, Lam. M. & G.
Sandy hillsides. Monongalia: near Ice's Ferry, and above Camp Eden. Upshur: near Buckhannon.

Var. **parviflora,** Gray.
Wet places. Gilmer: near Glenville—V. M.

L. inflata, L. Indian Tobacco, Lobelia. L. W. N. ,V. M., M. & G.
Dry soils. Common throughout the State.

Var. **simplex** (Raf.).
Dry places. Randolph: near Cricard, P. O. Characters of the species, but simple stemmed.
Having noted that this form perpetuated itself at one station in New York State, near Binghamton, for five years: I have decided that it is a true variety. Approaching the question from another point of view: I worked over a field near Morgantown this season, examining 783 small plants of L. inflata, many of which were not over four inches high, without finding a single simple-stemmed plant among them. At the station above named, as well as that in New York, there was a goodly amount of the variety, with none of the species in the immediate neighborhood.

SPECULARIA, Heist.

S. perfoliata, (L., A. DC. Venus' Looking-glass. M. & G.
Dry soils. Monongalia: near Morgantown. Upshur. near Buckhannon. Gilmer: near Glenville—V. M. Fayette: near Nuttallburg—L. W. N. Hardy: near Moorefield —G.

CAMPANULA. L.

C. rotundifolia, L. Harebell. M. & G.
Moist rocks. Mineral: along the Potomac, near Key-
ser—W. Gilmer: near Glenville—V. M. Tucker: along the
Blackwater.

C. aparinoides, Pursh. Marsh Bellflower.
Wet meadows. Preston: near Terra Alta.

C. Americana, L. Tall Bellflower. M. & G.
Rich woods, or even on dry rocks. Monongalia: near
Lee's Ferry. Wood, Wirt and Calhoun counties, general.
Gilmer: near Glenville— V. M., Prof. Brown. Lewis, and
Upshur. Randolph: near Cheat Bridge, alt. 3650 ft., with
wands 4-6 ft. high. Webster: in the glade region. Fayette:
near Nuttallburg—L. W. N.: along Loup Creek—James.
Kanawha and Jackson: general. Greenbrier: near White
Sulphur Springs. Summers: near Hinton. Marion: near
Worthington, and near Fairmont.

C. divaricata, Michx. M. & G.
Dry banks. Summers: near Talcott. Greenbrier:
near White Sulphur Springs.

VACCINIACEÆ.

GAYLUSSACIA, H. B. K.

G. dumosa (Andr), T. & G. Dwarf Huckleberry.
Damp, sandy soils. Kanawha: near Charleston—
James. Hardy: near Moorefield.

G. frondosa (L.), T. & G. Dangleberry.
Low copses. Fayette: near Hawk's Nest—James.
Webster: Upper Glade.

G. resinosa (Ait.). T. & G. Huckleberry.
Wirt: near Burning Springs. Monongalia: near
Laurel Hills. Marion: near Forksburg. Fayette: near
Nuttallburg—L. W. N. Frequent throughout the State.

OXYCOCCUS, Pers.

O. macrocarpus, Pers. Cranberry. (*Vaccinium macrocarpon,*
Ait.) M. & G.
Glades. Webster: Welsh, Long and Upper Glades.
(This station will be lost in a few years, as drainage is being
practised here to reclaim the land). Preston: Glade Farms,
Morgan's Glade, Cranberry, Reedsville and Terra Alta.

VACCINIUM, L.

V. stamineum, L.   Deerberry.]
　　Open woods. Wirt: near Burning Springs. Mineral:
along Knobby Mts.—W. Gilmer: near Glenville—V. M.
Fayette: near Nuttallburg—L. W. N. Grant: near Bayard.
Tucker: near Davis. Hardy: near Moorefield—G. Mercer:
near Bluefield.

V. Pennsylvanicum, Lam.   Dwarf Blueberry.
　　Dry hills. Gilmer: near Glenville—V. M. Brooke:
near Wellsburg—M. & G.

V. vacillans, Soland.   Low Blueberry.
　　Opens. Brooke: near Wellsburg M. & G. Fayette:
near Nuttallburg, alt. 2000 ft. -L. W. N.

V. corymbosum. L.   Swamp Blueberry. "Seedy Deerberry."
　　Swampy thickets. Preston: Kingwood Glades: Terra
Alta Glades. Webster: Welsh. Upper and Long Glades.

Var. **pallidum**. (Ait.). Gray.
　　Glady regions. Webster: in Upper Glade.

V. erythrocarpon. Michx.
　　Pocahontas: summit Spruce Knob, alt. 4800 ft.—A.
D. Hopkins.

CHIOGENES, Salisb.

C. hispidula (L.), T. & G. Creeping Snowberry.
　　Tucker: On rocks in the mist of **Blackwater Fall**.

ERICACEÆ.

GAULTHERIA, Kalm.

G. procumbens. L. Wintergreen. Tea-berry. Mountain Tea.
　　　　　　　　　　　　　　　　　　　　　　　　L. W. N.
　　Cool rich woods. Throughout the mountainous re-
gions of the State.

EPIGÆA, L.

E. repens. L. Trailing Arbutus.　　　　　　M. & G.
　　In moss of shady woods. Monongalia, and Preston:
along the Laurel Hills. Gilmer: near Glenville—V. M.
Mineral: near Keyser W. Kanawha: near Charleston—
James. Fayette: near Nuttallburg— L. W. N. Hardy: near
Moorefield—G. Mercer: near Bluefield.

ANDROMEDA, L.

A. ligustrina, Muhl.   "Seedy Buckberry."   M. & G.
Wet grounds. Preston: Morgan's Glade and Terra
Alta. Upshur: near Buckhannon. Webster: Upper, Long
and Welch Glades. Nicholas: Collett's Glade. Fayette:
near Nuttallburg—L. W. N.

Var. **pubescens**, Gray.
Swampy place. Fayette: near Nuttallburg, alt. 2000
feet; a variation with a six-lobed corolla and six-celled ovary
—L. W. N.

A. Mariana, L.   Stagger Bush.
Low grounds. Webster: Long Glade. Summers:
near Hinton.

OXYDENDRUM. DC.

O. arboreum, DC.   Sour Gum.   M. & G.
Rich woods. Wood: near Leachtown. Randolph:
near Valley Bend. Gilmer: near Glenville—V. M. Kanawha: near Charleston—James. Fayette: nor Nuttallburg
L. W. N. Summers: along Greenbrier River and near Hinton. Marion: near Shinnston and Clements. Monongalia:
near Beechwoods. Mercer: Beaver Spr. and Ingleside.

KALMIA, L.

K. latifolia, L.   Mountain Laurel. Calico-bush. Spoon-wood.
Dry or moist hillsides and thickets; forming impene-
trable masses in the mountains. Calhoun: Laurel Run.
Upshur: Sand Run. Webster: Buffalo Bull Mountains.
Kanawha: near Charleston—Barnes. Nicholas: near Beaver
Mills. Monongalia: near Ice's Ferry and Cheat View.
Preston: Laurel Hills. thence southward throughout the
eastern counties. Fayette: near Nuttallburg—L. W. N.
Jefferson: near Harper's Ferry—M. & G.

K. augustifolia, L...   Sheep-laurel. Lamb-kill.
Hillsides. Calhoun: Laurel Run. Upshur: Sand
Run. Nicholas: near Beaver Mills. Randolph: near Cheat
Bridge. Hardy: near Moorefield.

MENZIESIA, Smith.

M. globularis, Salisb.
Pocahontas: summit of Spruce Knob, alt. 4800 ft.
—A. D. Hopkins.

RHODODENDRON, L.

R. maximum, L. Great Laurel. James. L.W.N., V.M.. M.&G.
Deep rich woods, forming the most dense and tangled thickets in the mountains. Western limit on the Great Kanawha River near Charleston. Kanawha County. Common throughout the eastern and northern portions of the State.

R. arborescens, Torr. Smooth Azalea.
Glades and along mountain streams. Fayette: near Nuttallburg- L. W. N. Webster: Upper and Welch Glades.

R. canescens Michx.). Porter. Hoary Azalea.
Hampshire: Mutton Run, near Cacapon Springs; Dillon's Run, near Cacapon River.
Specimens in full leaf were noted at these points that differ widely from R. nudiflorum and R. calendulaceum, and seem, so far at least, to be this species.

R. viscosum (L.), Torr. M. & G.
Glades and cool ravines. Preston: Kingwood glades. Kanawha: near Charleston—Barnes. Fayette: near Nuttallburg—L. W. N.: near Hawk's nest—James. Webster: near Long Glade.

Var. **glaucum** (Lam.), Gray. "Cinnamon Honeysuckle."
Rocky streams in the higher mountains. Tucker: along the Blackwater Fork of Cheat.

Var. **nitidum** (Pursh,), Gray.
Glades. Webster: in Long and Upper Glades.

R. nudiflorum (L.), Torr. "Wild Honeysuckle." Pinxter Flower.
V. M., M. & G., W., L. W. N..
Rocky places along streams. Common throughout the northern, central, and eastern portions of the State. Mercer: near Princeton 6-8 ft. high.

R. calendulaceum (Michx.), Torr. Flaming Azalea.
Mountain woods. Monongalia: Cheat View. Mineral: near Keyser—W. Gilmer: near Glenville—V. M. Fayette: near Nuttallburg. alt. 1800 ft., flowers with a delicate fragrance quite distinct from that of other Azaleas—L. W. N. Summers: near Hinton. Preston: along the Laurel Hills. McDowell: near Elkhorn. Mercer: near Princeton and Bluefield.

R. Catawbiense. Michx. Lilac-colored Laurel.
Deep rich mountain woods, rare. Pendleton: near Cherry Grove. Fayette: near Nuttallburg, where it prefers

the face of cliffs—L. W. N. Greenbrier: Top of Alleghanies.
Summers: near Hinton.

CLETHRA, L.

C. acuminata, Michx. White Alder.
Wooded banks. Fayette: along the Gauley River at
the base of the Gauley Mountains: near Nuttallburg, uncommon—L. W. N.

PSEVA. Raf. (1809)
(Chimaphila, Pursh 1814.)

P. umbellata (L.). Prince's Pine.
Dry woods, rare compared with the next. Monongalia: along Decker's Creek: and on Laurel Hills in pine thickets.

P. maculata (L.). "Pipsisseway."   M. & G.
Rich woods, frequent throughout the northern, eastern, and central counties. Gilmer: near Glenville—V. M.
Prof. Brown. Kanawha: near Charleston—James. Fayette: near Hawk's Nest, and Kanawha Falls: near Nuttallburg—L. W. N. Hardy: near Moorefield—G.

MONESES. Salisb.

M. grandiflora, Salisb. One-flowered Pyrola.
Deep, cold woods. Gilmer: near Glenville—V. M.
Preston: along Laurel Hills. Monongalia: near Cheat View,
and along Quarry Run.

PYROLA, L.

P. elliptica, Nutt. Shin-leaf.   M. & G.
Rich woods. Kanawha: near Charleston—James.
Preston: near Terra Alta.

P. rotundifolia. L. Shinleaf.   M. & G.
Sandy woodlands, frequent. Upshur: summit on
Staunton Pike. Gilmer: near Glenville—V. M. Fayette:
near Nuttallburg—L. W. N. Grant: near Bayard. Hardy:
near Moorefield—G.

MONOTROPEÆ.

MONOTROPA, L

M. uniflora, L. Indian-pipe. Corpse-plant.   M. & G.
Deep, rich woods. Wirt: near Elizabeth. Kanawha:
near Charleston—James. Gilmer: near Glenville—V. M.

Fayette: near Nuttallburg—L. W. N. Preston: near Terra Alta. Grant: near Bayard. Hardy: near Moorefield—G.

HYPOPITYS, L.

H. Monotropa, Crantz. Pine Sap. (*Monotropa Hypopitys, L.*) Deep, rich woods. Wirt: near Elizabeth. Fayette: near Nuttallburg—L. W. N. Gilmer: near Glenville—V. M., Prof. Brown. Kanawha: near Charleston—James. Upshur: Summit on Staunton Pike.

DIAPENSIACEÆ.

GALAX, L.

G. aphylla, L. Woodlands. Fayette: near Nuttallburg—L. W. N. Monongalia: Dille Farm near Morgantown.

PRIMULACEÆ.

DODECATHEON L.

D. Meadia. L. Shooting Star. Rich woods. Mineral: near Keyser—W. Hardy: near Moorefield—G.

TRIENTALIS, L.

T. Americana (Pers.). Pursh. Star Flower. Damp cool woods. Mineral: near Piedmont. Grant: near Bayard. Tucker: along Blackwater Fork of Cheat. Preston: near Terra Alta. Monongalia: near Laurel Point.

STEIRONEMA, Raf.

S. ciliatum (L.). Raf.   M. & G. Low grounds and ditches. Randolph: on Rich Mountain, alt. 1610–2125 ft.; on Point Mountain. Grant: near Bayard. Gilmer: near Glenville Prof. Brown. Fayette: near Nuttallburg L. W. N.; near Hawk's Nest. Tucker: near Davis.

S. lanceolatum (Walt.). Gray. Low grounds. Wood: near Kanawha Station. Lewis: along Stone Coal Creek. Randolph: along Tygart's Valley River. Monongalia: Camp Eden.

Var. **angustifolium** (Lam.) Gray. Low grounds. Monongalia: Sandy banks of Cheat River above Camp Eden.

LYSIMACHIA L.

L. quadrifolia, L. Loosestrife. L. W. N., V. M., M. & G.
Moist soils. In all sections of the State visited.

L. terrestris (L.), B. S. P. (L. stricta, Ait.)
Wet places. Fayette: near Nuttailburg, in railroad
ditches, infrequent—L. W. N. Ohio: on Bogg's Island—M.
& G. Wood near Kanawha Station.

L. MUMMULARIA, L. Money-wort.
Escaped from cultivation. Wood: near Lockhart's
Run, profuse.

L. thyrsiflora. L. Pyramidal Loosestrife.
Wet meadows near the mountains. Upshur: near
Buckhannon. Randolph: along Tygart's Valley River.
Morgan: near Hancock.

ANAGALLIS, L.

A. ARVENSIS, L. Poor Man's Weather-glass.
Waste places. Jefferson: near Bolivar Heights—M.
& G.

SAMOLUS, L.

S. VALERANDI, var. **floribundus,** (H. B. K.), B.S. P.
var. Americanus, Gray.
Sandy places. Summers: shores of New River near
Hinton.

EBENACEÆ.

DIOSPYROS, L.

D. Virginiana, L. Persimmon. Date Plum. M. & G.
Thickets and opens. Wood: throughout. Fayette:
near Nuttallburg—L. W. N. Gilmer: near Glenville—V.
M. Monongalia: near Morgantown. Wirt: along Little
Kanawha River. Jackson: near Ripley. Lewis: along
Leading Creek.

STYRACACEÆ.

HALESIA, Ellis.

H. tetraptera, L. "Shittimwood."
Banks of streams. Fayette: near Nuttallburg—L. W.
N.: below Gauley Bridge. Summers: near Hinton, abun-
dant.

OLEACEÆ.

FRAXINUS. L.

F. Americana, L.   White Ash.   L. W. N., M. & G.
Rich woods. Frequent throughout the State.

F. pubescens, Lam.   Red or Black Ash.
Low grounds. Randolph: along Tygart's Valley
River. Upshur: near Lawrence. Fayette: near Nuttall-
burg—L. W. N. Wood: along the Little Kanawha River.
Mason: near Point Pleasant.

F. viridis, Michx. f.   Green Ash.
Along streams. Summers: near Hinton. Frequent
throughout the State.

F. sambucifolia, Lam.   Black Ash.
Wet woods. Wirt: along Straight Creek. Fayette:
near Nuttallburg, rare—L. W. N. Randolph: on Point
Mountain. Webster: Buffalo Bull Mountain. Monongalia:
near Ice's Ferry. Summers: Hinton.

CHIONANTHUS, L.

C. Virginica, L.   Fringe Tree.
River banks. Jackson: near Sandy and Ripley. Fay-
ette: near Nuttallburg, plentiful—L. W. N.: along Gauley
River near Gauley Mountains. Summers: near Hinton.
Monongalia: along Tibb's Run.

LIGUSTRUM. L.

L. VULGARE, L.   Privet.
Escaped from cultivation to waste places. Kanawha:
near Charleston—Barnes.

APOCYNACEÆ.

APOCYNUM, L.

A. androsæmifolium, L.   Spreading Dog's-bane.   M. & G.
Meadows, fields, and borders of thickets. Randolph:
near Cricard P. O.; near Valley Head. Greenbrier: near
White Sulphur Springs. Mason: near Point Pleasant; near
Brighton; near Buffalo.

A. cannabinum, L.   "Rheumatism weed." Indian Hemp. "Wild
Cotton." M. & G.
Moist grounds, fields, and banks of streams. Frequent
or common throughout the State.

Var. **pubescens** (R. Br.). DC.
Hardy: near Moorefield—G.

ASCLEPIAS, L.

A. tuberosa. L.   Pleurisy-root.   M. & G.
Fields and meadows. Wood: near Leachtown. Mo-
nongalia: near Morgantown and at Camp Eden. Lewis:
along Leading and Stone Coal Creeks. Webster: near Long
Glade. Fayette: near Nuttallburg—L. W. N.; near Kana-
wha Falls—James; near Gauley Bridge. Jackson: near
Fisher's Point. Gilmer: near Glenville—V. M.; Prof. Brown.
Doddridge: near Smithton. Jefferson: near Shenandoah
Jc. Berkeley: near Martinsburg. Hardy: near Moorefield
—G.

A. rubra. L.
Damp woods. McDowell: near Elkhorn.

A. purpurascens, L.   Purple Milk-weed.
Damp grounds. Fayette: near Nuttallburg—L. W.
N.; near Quinnimont. Hardy: near Moorefield—G.

A variegata. L.
Dry Woods. Wirt: near Elizabeth. Upshur: near
School House Summit. Hardy: near Moorefield—G. Fay-
ette: near Nuttallburg—L. W. N.

A. incarnata, L.   Swamp Milk-weed.   M. & G.
Wet places. Wirt: near Burning Springs. Gilmer:
along Tanner's Fork. Randolph: along Tygart's Valley
River, alt. 1963-2200 ft. Fayette: near Nuttallburg—L. W.
N. Monongalia: near Stewartown. Summers: near Hinton. Kanawha: near Charleston. Marion: near Worthington.

Var. **pulchra** (Ehrh.). Pers.
Hardy: near Moorefield—G.

A. Syriaca, L.   Milk-weed. *A. Cornuti.* Dec. G., V. M., M. & G.
Fields and roadsides. Common throughout the State,
even in the wildest portions.

A. exaltata (L.). Muhl. Poke Milk-weed. *A. phytolaccoides,* Pursh.
                                            M. & G.
Moist copses. Randolph: near Valley Bend; on Point
Mountain, alt. 1963-3300 ft. Preston: near Terra Alta.
Grant: near Bayard. Tucker: near Davis. Fayette: near
Nuttallburg—L. W. N.

A. quadrifolia. L. Four-leaved Milk-weed. M. & G.
Open woodlands. Mineral: near Keyser—W. Kanawha: near Charleston--Barnes. Gilmer: near Glenville—
V. M. Fayette: near Nuttallburg—L. W. N. Monongalia:
near Maple Run—M. H. Brown. Summers: near Hinton.
Hardy: near Moorefield—G.

ACERATES, Ell.

A viridiflora (Raf.). Ell. Green-flowered Milk-weed.
Berkeley: near Martinsburg—M. & G. Mineral: near
Keyser. Jefferson: near Charlestown. Jackson: along Limestone Ridge.

GONOLOBUS, Michx.

G. lœvis, Michx.
Climbing over weeds and fences. Mason: near Point
Pleasant. Putnam: near Buffalo.

LOGANIACEÆ.

MITREOLA, L.

M. petiolata. Torr. & Gray.
Damp soil. Fayette: near Nuttallburg: rare—L. W.
N.

GENTIANEÆ.

SABBATIA, Adans.

S. angularis (L.). Pursh. Centaury.
Rich soil. Jackson and Wood counties, frequent.
Gilmer: near Glenville—V. M. Fayette: near Nuttallburg
—L. W. N. Monongalia: near Easton. Summers: near
Hinton. Harrison: near Lumberport. Marion: near Clements.

GENTIANA, L.

G. quinquefolia. L. Four-leaved Gentian. (*G. quinqueflora.* Lam.)
Opens. Doddridge: near Long Run. Hardy: near
Moorefield—G.

G. Andrewsii, Grisch. Andrew's Gentian.
Moist woods. Fayette: near Nuttallburg. alt. 2000
ft.—L. W. N.. Monongalia: near Cheat View. Preston:
near Reedsville.

G. Saponaria, L.   Soapwort. Gentian.
Moist woods. Fayette: near Kanawha Falls—Selby.

G. linearis, Froel.
Boggy places. Preston: near Terra Alta and Morgan's Glade.

OBOLARIA, L.

O. Virginica, L.   Pennywort.
Moist woods. Gilmer: near Glenville—V. M. Fayette: near Nuttallburg—L. W. N. McDowell: near Elkhorn.

POLEMONIACEÆ.

PHLOX, L.

P. paniculata, Var. acuminata (Pursh.), Chapm.
Monroe: banks of Greenbrier river—L. W. N.

P. maculata, L.   Wild Sweet William.
Grassy woodlands along streams. Wirt: above Burning Springs, plentiful. Upshur: near Lorentz. Randolph: along Tygart's Valley River from Beverly to Valley Head. Hardy: near Moorefield—G.

P. amœna, Sims.
Dry open woods. Fayette: near Nuttallburg. frequent—L. W. N.

P. reptans, Michx.
Deep damp woods. Gilmer: near Glenville—V. M. Grant: near Bayard. Tucker: near Davis. Fayette: near Nuttallburg—L. W. N. Mercer: near Bluefield.

P. divaricata, L.   M. & G.
Rocky woods. Monongalia: near Morgantown and Stumptown. Gilmer: near Glenville—V. M., Prof. Brown. Hardy: near Moorefield—G. Fayette: near Nuttallburg—L. W. N. Mercer: near Bluefield.

P. sublata, L.   Moss Pink.
Dry rocky places. Mineral: near Keyser—W. Fayette: near Hawk's Nest—James. Monongalia and Marion:

near the F. M. & P. R. R. Hardy: near Moorefield—G. Mercer: near Bluefield.

POLEMONIUM, L.

P. reptans, L.   Greek Valerian.   M. & G.
Damp woods. Monongalia: in Brand's Woods near
Easton, where it is used by the people of that neighborhood
as a stomachic and tonic. Rich woods, near Morgantown.
Gilmer: near Glenville—V. M.

P. cœruleum, L.   Jacob's Ladder.
Moist opens. Preston: near Cranberry Summit.—M.
& G.

HYDROPHYLLACEÆ.

HYDROPHYLLUM, L.   '

H. macrophyllum, Nutt. Small-leaved Water-leaf.
Rich woods. Gilmer: near Glenville.—V. M. Wetzel:
near Burton.—M. & G.

H. Virginicum, L. Water-leaf.
Rich woods. Monongalia and Marion: along the
Monongahela river. Ohio: near Wheeling.—M. & G. Hamp-
shire: near Romney.

H. Canadense, L. Canadian Water-leaf.
Damp rich woods. Monongalia and Marion: along
the Monongahela River. Fayette: near Kanawha Falls—
James. Ohio: near Wheeling.—M. & G.

H. appendiculatum. Michx.
Ohio: Thomas' Hill, near Wheeling.—M. & G. Mineral: near Keyser—W. Grant: near Bayard. Tucker: near
the Falls of Blackwater.

PHACELIA, Juss.

P. bipinnatifida. Michx.
Rich soil. Fayette: near Nuttallburg, on or among
damp mossy rocks.—L. W. N.

P. Purshii, Buckley.
Moist wooded banks. Monongalia: banks of Decker's
Creek. Mineral: near Keyser.—W. Gilmer: near Glen-
ville.—V. M.

P. parviflora, Pursh.
Shaded banks. Fayette: near Nuttallburg—L. W. N.

ASPERIFOLIÆ.

CYNOGLOSSUM, L.

C. OFFICINALE, L. Hound's Tongue. "Dog-burr." M. & G.
Waste places and roadsides. Jefferson: near Shenandoah Jc. Gilmer: near Glenville—Prof. Brown. Hardy: near Moorefield—G. Mercer: generally frequent.

C. Virginicum. L. Wild Comfrey. M. & G.
Wood openings. Calhoun: near White Pine. Grant: near Bayard. Preston: near Terra Alta. Monongalia: near Morgantown Gilmer: near Glenville—V. M. Summers: near Hinton.

ECHINOSPERMUM, Sw.

E. Virginicum (L.), Lehm. Beggar's-lice. G., L. W. N., V. M.,
M. & G.
Borders and thickets. Frequent throughout the State.

MERTENSIA, Roth.

M. Virginica (L.), DC. Virginian Cowslip. Lung-wort. Blue-bells.
M. & G.
Rich woods. Monongalia and Marion: along the Monongahela River. Lewis: along Stone Coal Creek. Wirt: near Elizabeth. Upshur: near Laurentz. Gilmer: near Glenville—V. M., Prof. Brown. Hardy: near Moorefield —G.

ONOSMODIUM, Michx.

O. Carolinianum. DC.
Summers: banks of New River near Hinton.

MYOSOTIS, L.

M. PALUSTRIS (L.). Relh. Forget-me-not.
Damp places. Jefferson: near Harper's Ferry — M. & G.

SYMPHYTUM, L.

S. OFFICINALE, L. Comfrey.
Waste places. Gilmer: near Glenville—V. M. Mercer: near Ingleside.

LITHOSPERMUM, L.

L. ARVENSE, L. Corn Cromwell.
Fields. Ohio: near Wheeling—M. & G. Fayette: near Nuttallburg—L. W. N.

L. latifolium, Michx. Stone-seed.
Ohio: Cowan's Hill near Wheeling—M. & G. Fayette: near Nuttallburg; rare, not found in 1891—L. W. N. Monongalia: near Little Falls—K. D. Walker.

L. canescens (Michx.), Lehm. Puccoon.
Open woods. Mineral: on Knobby Mountain—W. Hardy: near Old Fields—A .D. Hopkins; and Moorefield—G. Hampshire: near Romney.

ECHIUM, L.

E. VULGARE, L. "Blue Weed." "Blue Devils." "Blue Thistle." "Blue Stem." M. & G.
Fields and waste ground. Jefferson: abundant especially near Charlestown, where there are many fields absolutely blue with the plant. Dr. Gray says of his trip through this country: "From the moment we entered the valley, we observed such immense quantities of Echium vulgare, that we were no longer surprised at the doubt expressed by Pursh whether it were really an introduced plant;" near Shenandoah Junction; Shepherdstown and Harper's Ferry. Randolph: along Tygart's Valley River; near Huttonsville, and up Riffles Creek. Berkeley: near Martinsburg, plentiful. Morgan: near Hancock, Cacapon and Orleans X Roads. Mineral: near Keyser, and Piedmont. Fayette: near Kanawha Falls—James; near Nuttallburg—L. W. N. Kanawha: opposite Coalburg. Summers: near Hinton. Jefferson: near Summit Point. Greenbrier: near White Sulphur Springs. Hardy: near Moorefield.

Also reported from: Jefferson: near Ripon. Summit Point, Middleway and Kabletown. Berkeley: near Oakton and Hedgesville. Morgan: near Rock Gap. Hampshire: near Slanesville, Concord, Romney, Three Churches. Dillon's Run. Higginsville. Sedan. Purgitsville and Springfield. Mineral: near Blaine. Hardy: near Moorefield. Wardensville and Old Fields. Grant: near Medley, Greenland and Petersburg. Tucker: near Hendricks and St. Georges. Pendleton: near Franklin and Upper Tract. Pocahontas: near Lobelia. Greenbrier: near Williamsburg and Fort Spring. Summers: near Talcott. Mercer: near Princeton and New Hope. McDowell: near Squire Jim. Wood: near Waverly. Lewis: near Vadis and Aberdeen. Barbour: near Old Field. Webster: near Replete. Wetzel: near Endicott. Doddridge:

near Smithton. Marshall: near Lowdenville. Roane: near Clio and Walnut Grove. Kanawha: near Tornado. Taylor: near Thornton; and Clay: near Valley Fork.

CONVOLVULACEÆ.

IPOMŒA, L.

COCCINEA, L. Scarlet Morning-glory.
Waste grounds. Monongalia: near Morgantown. Mason: near Point Pleasant.

HEDERACEA, Jacq. Ivy-leaved Morning-glory.
Waste places. Mason: sandy banks of the Ohio near Point Pleasant. Fayette: near Nuttallburg, banks of New River—L. W. N. Berkeley: near Martinsburg. Hardy: near Moorefield—G.

PURPUREA (L.). Lam. Morning-glory.
Fields, cultivated grounds and waste places. Mineral: near Keyser—W. Fayette: near Nuttallburg—L. W. N. Wood: near Lockhart's Run. Monongalia: near Morgantown. Jackson: near Sandyville. Mason and Putnam: an abundant weed in corn fields.

pandurata, (L.). Meyer. "Wild Sweet Potato." M. & G.
Fields, roadsides, and waste places. Monongalia: near Stewartown, Morgantown and Glenville. Marion: near Houghtown, Opekiska, and Fairmont. Wood: near Kanawha Station and Lockhart's Run. Gilmer: near Glenville—V. M. Fayette: near Nuttallburg—L. W. N. Greenbrier: near White Sulphur Springs. Monroe: near Alderson. Mason: near Point Pleasant. Summers: near Hinton. Hardy: near Moorefield—G.
Reported as a troublesome weed, from the following localities: Marion: near Canton, Farmington, Barracksville, Eldora, and Worthington. Taylor: near Grafton. Harrison: near Clarksburg, Bridgeport, Good Hope, Mt. Clair, and Wallace. Hampshire: near Slanesville, Concord, Three Churches, Bloomery, Dillon's Run, and Springfield. Jefferson: near Summit Point, Shenandoah Je., Middleway, and Kabletown. Jackson: near Douglass, Lone Cedar, Grass Lick, Garfield, Wilding, Odaville, Belgrove, and Kentuck. Ohio: near Alum Grove. Hancock: near New Cumberland. Lincoln: near Hamlin. Wood: near Waverly, Belleville, Tyner, Jerry's Run, Fountain Spring, Blennerhassett, Murphy's Mills, Deer Walk, and Rockport. Lewis: near Vadis, and Aberdeen. Wirt: near Burning Springs. Morris, Elizabeth, and Reedy Ripple. Summers: near Forest Hill, Talcott, Clayton, and Indian Mills. Preston: near Masontown,

and Reedsville. Wetzel: nerr Endicott, Pine Grove, New, Martinsville, and Blake. Mineral: near Patterson's Depot, and Piedmont. Berkeley: near Oakton, Martinsburg, and Gerardstown. Webster: near Replete. Ritchie: near Ritchie C. H., and Highland. Mercer: near Princeton. Concord Church. Bramwell and New Hope. Cabell: near Union Ridge and Milton. Kanawha: near Pocotaligo. Blandon and Gazil. Monroe: near Cashmere, and Johnson's X Roads. Wayne: near Adkin's Mills, and Stone Coal. Randolph: near Florence. Doddridge: near Smithton, Center Point, and Leopold. Fayette: near Fayetteville. Braxton: near Bulltown, Lloydsville, Frametown, and Newville. Tyler: near Wick, and Long Reach. Roane: near Newton, Looneyville, Clio, Reedy, Walnut Grove and Peneil. Upshur: near Evergreen, Kanawha Head, and Overhill. Barbour: near Pepper. Marshall: near Meighen, and Welcome. Grant: near Medley, and Greenland. Raleigh: near Egeira, and Raleigh C. H. Greenbrier: near Traut Valley, and White Sulphur Springs. McDowell: near Squire Jim. Mason: near Maggie. Taylor: near Thornton, and Meadland. Brooke: near Wellsburg, and Fowler's. Pleasants: near Schultz. Putnam: near Carpenter's. Hardy: near Old Fields. Clay: near Valley Fork.

I. lacunosa, L.

    Fayette: sandy banks of New River near Nuttallburg —L. W. N. Summers: near Hinton.

CONVOLVULUS. L.

C. spithamæus, L.

    Rocky soil. Mineral: near Keyser—W.

C. sepium. L. Hedge Bindweed.   M. & G

    Alluvial soils. Monongalia: near Little Falls: and along Decker's Creek. Mason: near Point Pleasant.

Var. **repens** (L.). Gray.

    Rocky river banks. Fayette: banks of New River near Nuttallburg—L. W. N. Monongalia: below Morgantown.

CUSCUTA, L.

C. TRIFOLII. Weihe. Clover Dodder.

    Parasitic on Clover. Greenbrier: near White Sulphur Springs.

C. Gronovii. Willd. Dodder.   G., L. W. N., M. & G

    Parasitic on grasses, sedges, and low weeds. Frequent in wet places throughout the State.

C. glomerata, Choisy.
Parasitic on Compositae. Monongalia: near Little Falls.

SOLANACEÆ.

SOLANUM, L.

S. DULCAMARA, L. Bitter-sweet. W., M. & G.
Damp places. Frequent throughout the State, but not so much so as the next.

S. nigrum. L. Common Nightshade. G., L. W. N.. V. M.. M. & G.
Fields, roadsides, and cultivated grounds. Common throughout the State.

S. CAROLINENSE, L. Horse Nettle. "Radical Weed." M. & G.
Becoming a detestable weed in fields and forests.
Calhoun: along Leading Creek. Wood: near Kanawha
Station. Wirt: near Elizabeth. Randolph: near Cricard P.
O. Webster: on Buffalo Bull Mountains. Nicholas: near
Beaver Mills. Gilmer: near Glenville-V. M. Fayette: near
Nuttallburgh-L. W. N. Monongalia: near Ice's Ferry.
Cabell: near Barboursville. Greenbrier: near White Sulphur
Springs. Monroe: near Alderson. Summers: near Hinton.
Kanawha: near Charleston. Mason: near Point Pleasant.
Jefferson: near Flowing Springs, and Shepherdstown. Mer-
cer: near Ingleside.
Reported as a troublesome weed from: Harrison:
near Clarksburgh, Wilsonburgh, Good Hope, Mt. Clair, and
Wallace. Ohio: near Elm Grove and West Liberty. Wood:
near Waverly, Belleville, Deer Walk and Kanawha Station.
Hardy: near Moorefield and Wardensville. Grant: near
Medley and Petersburg. Jefferson: near Moore's and
Kabletown. Summers: near Forest Hill and Talcott. Wetzel: near Endicott, Pine Grove, New Martinsville and Blake.
Mineral: near Patterson's Depot, and Blaine. Wirt: near
Burning Springs, Morris, Evelyn, and Reedy Ripple. Jackson: near Grass Lick, and Odaville. Cabell: near Union
Ridge, and Barboursville. Taylor: near Knottsville. Wayne:
near Stone Coal, and Adkin's Mills. Doddridge: near Smithton, and Center point. Marshall: near Knoxville, and Welcome. Braxton: near Bulltown, and Tate Creek. Berkeley:
near Hedgesville. Mercer: near Bramwell, and New Hope.
Roane: near Looneyville, Clio, Reedy, and Pencil. Pocahontas: near Lobelia. Kanawha: near Blandon. Greenbrier:
near Trout Creek. McDowell: near Squire Jim. Mason:
near Maggie. Brooke: near Wellsburgh. Marion: near
Mannington. Taylor: near Grafton. Upshur: near Kanawha Head, Overhill, and Hemlock. Hampshire: near

Higginsville and Springfield. Tyler: near Long Reach.
Webster: near Welsh Glade. Clay: near Valley Fork.

PHYSALIS, L.

P. Philadelphica, Lam.
Rich opens. Fayette: near Nuttallburg—L. W. N.

P. angulata. L. Ground Cherry.
Open rich grounds. Grant: near Bayard. Gilmer:
near Glenville—V. M. Wood: near Kanawha Station.

P. pubescens, L.
Low grounds. Fayette: near Nuttallburg—L. W. N.
Ohio: near Wheeling—M. & G. Jefferson: near Shepherds-
town. Mason: near Point Pleasant.

P. Virginiana, Mill.
Light sandy soils. Monongalia: near Beechwoods.
Jefferson: near Shepherdstown. Jackson: near Ripley.
Hardy: near Moorefield—G.

P. viscosa, L.
Low grounds. Ohio: near Wheeling—M. & G. Hardy:
near Moorefield—G.

P. lanceolata, Michx.
Dry opens. Jackson: near Ripley. Wood: near
Sharktown.

PHYSALODES, Bohn, (1760)
(Nicandra, Adans, 1763)

P. PHYSALOIDES (L.) Apple-of-Peru.
Waste grounds. Lewis: near Weston. Mineral: near
Piedmont.

LYCIUM, L.

L. VULGARE (Ait), Dun. Matrimony Vine.
A frequent and persistent escape. Berkeley: near
Martinsburgh. Jefferson: near Shepherdstown. Mason:
banks of the Ohio near Point Pleasant.

DATURA, L.

D. STRAMONIUM L. "Jimson-weed." Jamestown-weed. Stink-
weed. L. W. N., V. M., M. & G., G.
Waste places. A common weed throughout the State.

D. TATULA. L. Purple Thorn-apple. L. W. N., V. M., M. & G.
With the last. Common throughout the State.

PETUNIA, L.

P. VIOLACEA, Hook.
Escaped to waste grounds. Monongalia: near Morgantown, common, where it persists annually. Mason: near Point Pleasant. Jefferson: near Shepherdstown.

SCROPHULARINEÆ.

VERBASCUM, L.

V. THAPSUS. L. Mullein. L. W. N., V. M., M. & G.
Old fields and pastures. Common throughout the State.

V. BLATTARIA, L. Moth Mullein. M. & G.
Fields and waste places. Wirt: along the Little Kanawha River. Gilmer: near Glenville—V. M. Fayette: near Nuttallburg—L. W. N. Monongalia and Marion, common. Grant: near Bayard. Jefferson: near Charlestown. Berkeley: near North Mountain. Elsewhere frequent.

V. LYCHNITIS, L. Yellow Moth-mullein.
Fields and wastes. Kanawha: roadside, up 8 Mile Creek. Mineral: opposite Cumberland, Md.

LINARIA, Juss.

L. VULGARIS, Mill. Toad Flax. "Devil's Flax." "Wild Flax." "Indian Hemp." "Impudent Lawyer." M. & G.
Fields, roadsides and waste places. Monongalia: near Stewarttown. Jefferson: near Charlestown and Shenandoah Junction. Jackson: near Sandyville. Berkeley: near North Mountain. Mineral:, near Piedmont and Keyser. Hardy: near Moorefield—G.
Also reported as a weed from: Harrison: near Good Hope. Ohio: near West Liberty. Wood: near Jerry's Run. Fountain Spring, Blennerhassett, and Rockport. Jefferson: near Molers. Wetzel: near Endicott. Mineral: near Piedmont. Wirt: near Burning Spring. Morris and Reedy Ripple, Jackson: near Lone Cedar, Garfield, and Belgrove. Cabell: near Union Ridge. Taylor: near Thornton and Meadland. Wayne: near Stone Coal. Marshall: near St. Joseph. Berkeley: near Martinsburg. Greenbrier, near White Sulphur Springs. Mason: near Grimm's Landing and Maggie. Upshur: near Kanawha Head and Overhill. Tyler: near Long Reach. Webster: near Welch Glade.

Preston: near Tunnelton and Terra Alta. Hancock: near New Cumberland and Fairview. Ritchie: near Ritchie C. H. Monroe: near Pickaway. Morgan: near Rock Gap, Tucker: near Texas. Raleigh: near Raleigh C. H.

SCROPHULARIA, L.

S. nodosa, L., *var.* **Marilandica** L., Gray. Figwort. M. & G.
Fields and waste places. Monongala: near Morgantown. Fayette: near Nuttallburg L. W. N. Greenbrier: near White Sulphur Springs. Kanawha: near Charleston.

COLLINSIA, Nutt.

C. verna, Nutt.
Moist soil. Gilmer: near Glenville—V. M.; Prof. Brown. Ohio: on Wheeling Hill—M. & G. Monongalia: near Cassville.

CHELONE, L.

C. glabra, L. Snake Head.
Wet places. Gilmer: near Glenville V. M Fayette: near Nuttallburg. uncommon —L. W. N. Wirt: near Burning Spring.

C. obliqua. L.
Wet places. Randolph: along Cheat River near Cheat Bridge. Monongalia: ner Camp Eden.

PENTSTEMMON. Mitch.

P. hirsutus L), Willd. Beard-tongue.   (P. *pubescens*, Sol.)
Dry or rocky places Kanawha: near Charleston—Barnes. Gilmer: near Glenville—V. M. Cabell: near Barboursville—James. Hardy: near Moorefield—G. Hampshire: near Doe's Gully.

P. lævigatus, Soland.
Rich soil. Fayette: near Nuttallburg. in open woods L. W. N.

Var. **Digitalis**(Nutt.) Gray.
Rich soil. Monongalia: along the Monongahela River, frequent. Wood: near Kanawha Station. Fayette: near Kanawha Falls—James. Hardy: near Moorefield. Hampshire: near Doe's Gully.

MIMULUS, L.

M. ringens. L. Monkey Flower.   M. & G.
Wet places. Upshur: along Stone Coal Creek. Fayette: near Kanawha Falls—James; near Nuttallburg— L. W. N. Randolph; along Tygart's Valley River. Frequent: throughout the State.

M. alatus. Ait. Winged Monkey Flower.
Wet places. Greenbrier: near White Sulphur Springs. Putnam: near Buffalo.

GRATIOLA, L.

G. Virginiana, L. Gratiola.   L. W. N., M. & G.
Ditches. Common throughout the State.

G. sphærocarpa, Ell.
Damp places. Fayette: near Nuttallburg, on mossy banks in bed of creek.—L. W. N.

ILYSANTHES, Raf.

I. gratioloides(L.), Benth.   False Pimpernel.   (*I. riparia*, Raf.)
Wet places. Fayette: near Kanawha Falls—James. Along Little Kanawha River—M. & G.

VERONICA, L.

V. Virginica, L.   Culver's Physic.
Rich woods and borders. Webster: Long Glade. Jackson: near Sandyville. Fayette: near Nuttallburg.— L. W. N.

V. Anagallis, L.   Water Speedwell.
Banks and ditches. Fayette: near Kanawha Falls— James.

V. Americana, Schw. American Brooklime. "Wallink." M. & G.
Brooks and ditches. Monongalia: the Flats near Morgantown. Randolph: on Point Mountain, alt. 3,050 ft. (where it is called "Wallink," and is used internally to bring out rashes.) Webster: near Addison, alt. 2,000 ft. Mercer: near Beaver Spr.

V. officinalis, L.   Speedwell. "Gypsy Weed."   M. & G.
Rich, deep woods and opens. Randolph: on Rich Mountains. Gilmer: near Glenville—V. M. Kanawha: near Charleston—James. Fayette: near Nuttallburg—L. W. N. Monongalia: along Decker's Creek. Greenbrier: near

White Sulphur Springs; and frequent throughout the State.
Mercer: near Beaver Spr., and Bluefield.

V. serpyllifolia, L. Thyme-leaved Speedwell.
Roadsides, fields and lawns. Monongalia: near Morgantown. Gilmer: near Glenville—V. M. Fayette: near Nuttallburg—L. W. N. Ohio: near Wheeling—M. & G.
Mercer: near Bluefield.

V. peregrina, L. Neck Weed. Purslane-speedwell.
Waste places. Monongalia: near Morgantown. Fayette: near Nuttallburg—L. W. N. Ohio: Thomas Hill.
near Wheeling—M. & G.

V. ARVENSIS. L. Corn Speedwell.
Cultivated grounds. Gilmer: near Glenville—V. M.
Ohio: Thomas Hill, near Wheeling—M. & G. Fayette: near Nuttallburg—L. W. N.

BUCHNERA, L.

B. Americana, L. Blue Hearts.
Moist, sandy ground. Putnam: near Buffalo.

GERARDIA, L.
G. pedicularia, L.
Hardy: near Moorefield—G. Fayette: near Nuttallburg—L. W. N.

G. tenuifolia, Vahl. Slender Gerardia.
Dry soil. Mineral: near Keyser—W. Fayette: near Nuttallburg, alt. 2000 ft.—L. W. N. Randolph: near Elkins.

G. Virginica L. J. B.S.P. Oak-leaved Gerardia. *G. quercifolia*, Pursh.
Open woods. Fayette: near Nuttallburg—L. W. N.

G. flava. L. False Foxglove.
Open woods. Wood: near Leachtown. Fayette: near Kanawha Falls and Hawk's Nest—James: near Nuttallburg—L. W. N. Gilmer: near Glenville—V. M. Hardy: near Moorefield—G.

G lævigata. Raf.
Oak woods. Fayette: near Nuttallburg, alt. 2000 ft.
—L. W. N.

G auriculata, Michx.
Fields. Mongalia: near Little Falls, with pure white flowers.

CASTILLEJA, L. f.

C. coccinea(L.), Spreng. Painted Cup.
Sandy soils. Monongalia: along the Monongahela at
Uffington, and below Morgantown. Randolph: near Valley
Head. Preston: near Terra Alta. Hampshire: near Rom-
ney.

PEDICULARIS, L.

P. Canadensis, L. Louse-wort. M. & G.
Copses, woods and banks. Preston: near Terra Alta.
Fayette: near Nuttallburg—L. W. N. Gilmer: near Glen-
ville—V. M. Greenbrier: near White Sulphur Springs.
Summers: near Hinton. McDowell: near Elkhorn.

MELAMPYRUM, L.

M. lineare, Lam. Cow-wheat. *M. Americanum.* Michx.
Rich, open woods. Preston: near Terra Alta.

OROBANCHACEÆ.

EPIPHEGUS, Nutt.

E. Virginiana(L), Bart. Beech-drops. Cancer-root.
Parasitic upon the roots of the beech. Wirt: near
Elizabeth. Gilmer: near Glenville—V. M. Fayette: near
Nuttallburg—L. W. N. Monongalia: near Morgantown.

CONOPHOLIS, Wallr.

C. Americana(L.f.), Wallr. Cancer Root.
Oak woods. Among fallen leaves. Monongalia:
along Decker's Creek and near Little Falls. Gilmer: near
Glenville—V. M. Fayette: near Nuttallburg—L. W. N.
McDowell: near Elkhorn. Mercer: near Bluefield.

APHYLLON, Mitch.

A. uniflorum(L.), Gray. One-flowered Cancer-root. M. & G.
Damp woodlands and opens. Monongalia: near Mor-
gantown. Wirt: near Elizabeth. Gilmer: near Glenville—
V. M. Fayette: near Nuttallburg—L. W. N.

BIGNONIACEÆ.

TECOMA. Juss.

T. radicans(L.). Juss. Trumpet-Creeper. M. & G.
Moist soil. Monongalia: near Little Falls. Mari n:
near Fairmont. Fayette: near Nuttallburg—L. W. N.
Summers: near Hinton. Mason: near Point Pleasant; and
frequent throughout the State.

CATALPA. Juss.

C. Bignonioides. Walt. Indian Bean Tree. "Catawba." M. & G.
River banks. Marion: near Catawba, a place named
after this tree, which would render it apparent that the tree
was native here, which I hold to be true. Wood: near
Leachtown. Gilmer: near DeKalb and along Leading Creek.
Monongalia: near Lee's Ferry and Stewartown. Wirt: near
Elizabeth. Upshur: along Big Sandy Run, alt. 1827 ft.
Mason: near Point Pleasant.

ACANTHACEÆ.

RUELLIA. L.

R. ciliosa, Pursh.
Margins of woods. Wood: near Leachtown. Wirt:
near Elizabeth. Calhoun: near Grantsville. Gilmer: near
DeKalb. Lewis: along Stone Coal Creek. Upshur: near
Laurentz. Nicholas: along the Gauley River. Kanawha:
near Cannellton.

DIANTHERA, L.

D. Americana, L. Water Willow. L. W. N.. M. & G.
In streams. Common throughout the State.

VERBENACEÆ.

VERBENA, L.

V. OFFICINALIS, L. Vervian.
Jefferson: near Harper's Ferry—M. & G.

V. urticæfolia, L. White Vervian. L. W. N., M. & G.
Waste or open grounds. Common throughout the
State.

V. haɪtata, L. Blue Vervian.   L. W. N.
Damp waste grounds and roadsides. Frequent
throughout the State. In some places rare.

V. angustifolia. Michx.   M. & G.
Roadsides and waste places. Throughout Jefferson
County. Hardy: near Moorefield—G.

LIPPIA, L.

L. lanceolata, Michx. Frog Fruit.
Waste grounds. Ohio: near West Wheeling—M. &
G. Mason: Banks of the Ohio River near Point Pleasant.

PHRYMA, L.

P. Leptostachya, L. Lop-seed.   M. & G.
Rich open woods. Greenbrier: near White Sulphur
Springs. Fayette: near Kanawha Falls—James: near Nuttallburg—L. W. N. Kanawha: near Charleston. Mason:
near Point Pleasant: and frequent throughout the State.

LABIATÆ.

ISANTHUS, Michx.

I. brachiatus(L.). B. S. P. False Pennyroyal. *I, cæruleus*, Michx.
Mineral: opposite Cumberland, Md.—M. & G.

TEUCRIUM, L.

T. Canadense, L. Germander. Wood Sage.
Low grounds. Wood: ditches near Kanawha Station.
Jackson: near Sandyville. Gilmer: near Glenville—V. M.
Cabell: near Barboursville—James. Fayette: near Nuttall-
burg, profile of expanded flower resembles a deer's head—L.
W. N. Monongalia: near Ice's Ferry. Hardy: near Moore-
field—G.

COLLINSONIA. L.

C. Canadensis. L. Rich-weed. Stone-root.
Rich, damp woods. Randolph: along Staunton pike
up Riffles Creek, alt 2700 ft. Gilmer: near Glenville—V.
M. Fayette: near Nuttallburg, some plants with elliptical
leaves, acute at both ends 10 in. wide by 3½ inches long—
L. W. N. Monongalia: near Camp Eden. McDowell: near
Elkhorn.

PERILLA, L.

P. OCYMOIDES. L.. *rar.* **CRISPA.**
Waste ground. Monongalia: near Morgantown.

MENTHA, L.

M. VIRIDIS, L. Spearmint.   G.. V. M., M. & G.
Low grounds and damp places. Frequent, even at
the higher altitudes.

M. PIPERITA, L.. Peppermint.   M. & G.
A frequent escape along springy brooklets. Gilmer:
along Tanner's Fork. Randolph: on Point Mountain, alt.
3050 ft. Jackson: near Sandyville. Gilmer: near Glenville—V. M. Fayette: near Nuttallburg, alt. 2000 ft—L. W.
N. Summers: near Hinton.

M. SATIVA, L. Whorled Mint.
Monroe: banks of Greenbrier River—L. W. N.

M. Canadensis. L. Wild Mint.
Wet places. Randolph: along the road up Point
Mountain. alt., 2325 feet. Fayette: near Nuttallburg, rare
—L. W. N. Summers: near Hinton.

LYCOPUS, L.

L. Virginicus, L.. Bugle Weed.   L. W. N.
Low, wet grounds. Common throughout the State.

L. sinuatus, Ell.
Low. wet ground. Mason: near Point Pleasant.

CUNILA, L.

C. Mariana. L. Dittany.
Dry hillsides. Gilmer: near Glenville—V. M. Fay-
ette: near Nuttallburg—L. W. F. Wetzel: near Burton—
M. & G.

KOELLIA, Moen. (1794.)
(Pycanthemum, Michx. 1803.)

K flexuosa Walt.).   (*P. linifolium,* Pursh.)
Dry grounds. Wood: near Kanawha Station and
Lockhart's Run. Fayette: near Kanawha Falls James:
near Nuttallburg—L. W. N. Preston: near Terra Alta.
Summers: near Hinton. Monroe: near Alderson.

425

K. Torreyi(Benth.).
    Dry soil. Fayette: near Nuttallburg—L. W. N.
Summers: near Hinton.

K. clinopodioides(T. & G.)
    Dry soil. Nicholas: near Beaver Mills. alt. 2125 ft.

K. Tullia(Benth.).
    Open woods and banks. Fayette: near Nuttallburg.
common—L. W. N.

K. incana(L.). Mountain Mint.
    Dry soils. Wirt: near Burning Springs and Elizabeth.
Monongalia: near Morgantown. Fayette: near Nuttallburg
—L. W. N.

K. montana(Michx.).
    Rocky river banks. Fayette: near Nuttallburg. rare.
Not found in 1891—L. W. N.

HEDEOMA, Pers.

H. pulegioides(L.). Pers. American Pennyroyal. L. W. N.,
                                         M. & G.
    Dry fields and woods. Common throughout the
State.

CALAMINTHA, Mœnch.

C. Clinopodium, Benth. Basil.   M. & G.
    Dry soils. Upshur: near Buckhannon. Fayette:
near Nuttallburg. rare—L. W. N. Randolph: near Crickard
P. O.

MELISSA, L.

M. OFFICINALIS, L. Balm.
    Escaped from gardens. Kanawha: up 8-Mile Creek.
Fayette: near Nuttallburg—L. W. N.

SALVIA, L.

M. lyrata, L. Sage.
    Meadows. Monongalia: near Morgantown. Fayette:
near Nuttallburg L. W. N. Gilmer: near Glenville—V. M.
Mercer: near Ingleside.

MONARDA. L.

M. didyma, L. Bee-balm. Oswego-Tea.   M. & G.
    Moist places. Randolph: near Cheat Bridge, alt. 3350

ft.; near Valley Head. Mineral: near Davis. Grant and Tucker on W. Va. Central R. R. Monroe: near Alderson. Hardy: near Moorefield—G.

M. fistulosa, L.. Wild Bergamont. M. & G.
Dry soils. Wirt: near Elizabeth. Gilmer: near De-Kalb, abundant: near Glenville- V. M. Randolph: summit of Point Mountain. alt. 3700 ft. Monongalia: near Ice's Ferry. Fayette: near Nuttallburg—L. W. N.; near Kanawha Falls—James. Kanawha: near Coalburg—James. Greenbrier: near White Sulphur Springs. Marion: near Fairmont. Hardy: near Moorefield- G.

Var. **rubra,** Gray.
Moist grounds. Mineral: along Abraham's Creek. Summers: near Greenbrier Stockyards. Monroe: near Alderson.

Var. **mollis**(L.., Benth.
Shady places. Fayette: near Nuttallburg—L. W. N.

BLEPHILIA, Raf.

B. hirsuta. Benth. M. & G.
Fields and fence rows. Randolph: summit of Rich Mountain, alt. 3000 ft. Fayette: near Hawk's Nest—James. Preston: near Terra Alta.

AGASTACHE, Gron. (1762).
(Lophanthus, Benth, 1834.)

A. nepetoides(L.) Giant Hyssop.
Ohio: near Wheeling—M. & G.

CEDRONELLA, Moench.

C cordata. Benth.
Moist, shady ravines. Kanawha: near Charleston—Barnes. Fayette· near Kanawha Falls—James. Gilmer: near Glenville- V. M.; Prof. Brown. Randolph: summit of Point Mountain. alt. 3700 ft. Monongalia: near Round Bottoms; opposite Little Falls. Fayette: near Nuttallburg—L. W. N.

NEPETA, L.

N. CATARIA. L. Catnip. L. W. N.. V. M., M. & G.
Roadsides and waste places. Common throughout the State.
Found at various points in the higher Alleghanies, remote from dwellings.

N. HEDERACEA(L.). B. S. P. Ground Ivy. Gill-over-the-ground.
N. Glechoma. Benth. L. W. N.,M. & G.
Abundant throughout the settled portions of the State.

SCUTELLARIA, L.

S. lateriflora. L. Mad-dog Skull-cap. L. W. N., M. & G.
Wet shady places. Frequent throughout the State.

S. versicolor, Nutt. *var.* **minor.** Chapm.
Rich soil. Fayette, near Nuttallburg. L. W. N. On
visiting Mr. Nuttall's station for this species, a moss covered
boulder, I was impressed with the great beauty of this little
skull-cap, which, in its mossy bed, resembled a bright blue
bit of color upon a Fairy's palette.

S. saxatalis, Riddell. M. & G.
Moist shady banks. Fayette: near Nuttallburg—L.
W. N.; along the north bank of the Great Kanawha River
near Kanawha Falls.

S. serrata. Andrews.
Woodlands: Kanawha: near Charleston—Barnes.
Putnam: near Buffalo. Fayette: near Nuttallburg—L. W. N.

S. canescens, Nutt. M. & G.
Ditches and moist places. Wirt: near Elizabeth.
Kanawha: up 8 Mile Creek.

S. pilosa, L.
Dry mountain sides. Fayette: near Nuttallburg—
L. W. N.

Var. **hirsuta,** Benth.
With the preceding. Fayette: near Nuttallburg—
L. W. N.

S. integrifolia, L., *var.* **hyssopifolia.**
Low grounds. Wood: near Kanawha Station,
abundant.

S. parvula. Michx.
Sandy banks. Wood: near Parkersburg—M. & G.
Mason: near Point Pleasant.

S. galericulata, L.
Wet shady places. Kanawha: near Charleston
Barnes.: near Pocataligo. Jackson: near Fisher's Point.
Gilmer: near Glenville—Prof. Brown.

Forma **albiflora.**
> Kanawha: near Charleston—Barnes.

S. nervosa, Pursh.   M. & G.
> Moist thickets. Monongalia: on The Flats near Morgantown.

BRUNELLA, L.

B. vulgaris, L. Heal-all.   G., L. W. X., V. M., M. & G.
> All situations. Common throughout the State.

Forma **albiflora**(Boggenhard), Britt.
> Jackson: on Limestone Ridge.

PHYSOSTEGIA. Benth.

P. Virginiana(L.). Benth. False Dragon-head.
> Wet places. Fayette: near Nuttallburg—L. W. X.;
> near Kanawha Falls—James.

MARRUBIUM, L.

M. VULGARE, L. Horehound.   M. & G.
> Waste grounds, escaped from gardens. Randolph:
> near Ford's. Jefferson: near Shepherdstown. plentiful.

STACHYS, L.

S. palustris, L. Hedge Nettle.
> Wet grounds. Gilmer: near Glenville—Prof. Brown.

S. aspera, Michx.
> Damp places. Fayette: near Nuttallburg—L. W. X.

Var. **glabra,** Gray.
> Damp places. Mason, banks of the Ohio River near
> Point Pleasant. common.

S. cordata, Ridd.
> Rocky thickets. Wirt: near Elizabeth.

GALEOPSIS, L.

G. TETRAHIT, L. Hemp Nettle.
> Waste places. Preston: near Terra Alta.: near Cranberry Summit. M. & G.

LEONURUS. L.

L. CARDIACA, L.. Motherwort
Waste places near dwellings. Monongalia: near Lee's
Ferry. Hardy: near Moorefield. Mercer: near Princeton.
Jefferson: near Shenandoah Je.

LAMIUM L.

L. AMPLEXICAULE, L.. Dead Nettle.
Escaped from Gardens. Fayette: near Nuttallburg-
L. W. N. Monongalia: plentiful on the College Campus.

TRICHOSTEMA, L.

T. dichotomum, L.. Bastard Pennyroyal.
Dry fields. Mason: near Brighton. Hardy: near
Moorefield—G.

PLANTAGINEÆ.

PLANTAGO, L.

P. MAJOR, L.. Plantain.
Waste ground. Ohio: near Wheeling—M. & G.
Fayette: Nuttallburg—L. W. N. Monongalia: Morgantown.

P RUGELII, Decne. Common Plantain.   L. W. N., M. & G.
Common throughout the State, near dwellings.

P. LANCEOLATA, L. "Buck Plantain." "Ripple." "Buck-horn
Plantain."
Becoming a common weed throughout the State;
very little as yet however in Jefferson. Berkeley and Morgan
counties. Greenbrier: near White Sulphur Springs. Fayette: near Nuttallburg—L. W. N. Mercer: near Princeton.

P. Virginica, L. White Plantain.   M. & G.
Sandy soils. Fayette: near Nuttallburg, L. W. N.
Monongalia: near Morgantown: and frequent throughout
the State. Hardy: near Moorefield.

APETALÆ.

ILLECEBRACEÆ.

ANYCHIA. Rich.

A. Candensis L.). B. S. P. (*A. capillacea*, D. C.)
Dry Woods. Fayette: near Nuttallburg. common—
L. W. N.

PARONYCHIA, Tourn.

P. dichotoma, Nutt. Whitlow-wort.
Rocky places. Jefferson: near Harper's Ferry—Gray.

AMARANTACEÆ.

AMARANTUS. L.

A. HYPOCHONDRIACUS, L. Red Amaranth.
Waste places. Ohio: on Bogg's Island—M. & G.
Monongalia: near Morgantown.

A. PANICULATUS, L.
Waste places. Monongalia: near Morganton. Hardy: near Moorefield—G.

A. RETROFLEXUS, L. Pigweed.
Ohio: on Bogg's Island—M. & G. Monongalia: near Morgantown.

A. CHLOROSTACHYS, Willd.
Gardens and waste places. Fayette: near Nuttallburg. common—L. W. N. Monongalia: near Morgantown. Hardy: near Moorefield—G.

A. ALBUS, L. Tumble-weed. M. & G.
Waste places. Monongalia: near Morgantown.

A. SPINOSUS, L. Thorny Amaranth.
Waste grounds. Kanawha: near Charleston. Putnam: near Buffalo. Mason: near Point Pleasant. Wood: near Parkersburg. abundant. Jefferson: near Shepherdstown.

431

CHENOPODIACEÆ.

CHENOPODIUM, L.

C. ALBUM, L.* Lamb's Quarters. Pigweed. L. W. N., M. & G.
Roadsides and waste places, common throughout the
State.

Var. **VIRIDE,** Moq.
Dry sandy hillsides. Fayette: near Nuttallburg, alt.
2000 ft., uncommon—L. W. N.

C. HYBRIDUM, L.
Dry sandy hillsides. Fayette: near Nuttallburg, apparently indigenous—L. W. N. Jefferson: near Shepherdstown.

C. URBICUM, L. M. & G.
Waste places, frequent.

C. GLAUCUM, L.
Waste places. Monongalia: near Ice's Ferry.

C. BOTRYS, L. Jerusalem Oak.
Ohio: near Wheeling. Jefferson: Shepherdstown.

C. AMBROSIOIDES, L. Mexican Tea. M. & G.
Waste places. Common. Kanawha: along Great
Kanawha River. Taylor: near Grafton. Wood: near
Parkersburg.

Var. **ANTHELMINTICUM**(L.), Gray. Worm weed.
Plentiful along the Great Kanawha River. In Kanawha, Putnam and Mason counties. Fayette: near Nuttallburg—L. W. N. Jackson: along C. & P. pike.

PHYTOLACCACEÆ.

PHYTOLACCA. L.

P. decandra, L. Poke. Scoke. Garget. G., L. W. N., M. & G.
All situations, even in higher mountains. Common
throughout the State.

POLYGONACEÆ.

ERIGONUM. Michx.

E. Alleni, Wats.
Greenbrier: near White Sulphur Springs—T. F. Allen.

POLYGONUM. L.

P. ORIENTALE. L. Prince's Feather M. &. G.
Escaped to waste places. Lewis : near Weston. Mon-
ongalia: near Morgantown. Fayette: near Nuttallburg, rare
—L. W. N. Mineral: near Piedmont.

P. Pennsylvanicum, L. G., L. W. N.
Low grounds. Common throughout the State.

P. PERSICARIA, L. Lady's Thumb. M. & G.
Waste grounds. Lewis: near Weston. Gilmer: near
Glenville—V. M. Fayette: near Nuttallburg—L. W. N.
Mason: near Point Pleasant. Wood: near Parkersburg.
Hardy: near Moorefield—G.

Forma **albiflora.**
A pure white-flowered form abundant near Point
Pleasant.

P. Hydropiper, L. Smartweed. Water Pepper.
Wet grounds. Monongalia: near Morgantown. Gil-
mer: near Glenville—V. M. Fayette: near Nuttallburg—L.
W. N. Ohio: near Wheeling—M. & G.

P. acre, H. B. K. G., L. W. N.
Wet places. Common throughout the State.

P. hydropiperoides, Michx. Mild Water-pepper.
Swampy places. Common throughout the State.

P. Virginianum, L. L. W. N.
Thickets and in rich soils. Common throughout the
State.

P. aviculare, L. Door-weed. "Goose-grass." L. W. N.
About dwellings and roadsides. Common throughout
the State.

P. erectum. L. L. W. N
With the last, especially in streets.

P. tenue. Michx.
Dry pastures. Wood: near Kanawha Station. Monongalia: near Morgantown. Taylor: near Grafton, and
common throughout the State.

P. sagittatum, L. Tear-thumb. G., L. W. N., M. & G.
Wet places. Common throughout the State.

P. arifolium, L.
Low grounds. Berkeley: near Martinsburg. Monongalia: near Morgantown, and elsewhere frequent.

P. CONVOLVULUS, L. Black Bindweed.
Gardens and waste places. Fayette: near Nuttallburg —L. W. N. Mason: near Point Pleasant.

P. dumetorum. L... *var.* scandens(L.)Gray. Climbing False Buckwheat. G., L. W. N., M. & G.
Low grounds along streams.

FAGOPYRUM, Gærtn.

F. ESCULENTUM, Moench. Buckwheat. G. W. N., M. & G.
Waste grounds and cultivated fields. A frequent escape.

RUMEX, L.

R. Brittanicus, L. Water Dock.
Wet places. Ohio: on Bogg's Island—M. & G. Berkeley, near Martinsburg.

R. CRISPUS. L. Curled Dock. M. & G.
Waste places and cultivated fields. Frequent.

R. OBTUSIFOLIUS. L. Bitter Dock. L. W. N.
Waste grounds and cultivated fields, common.

R. CRISPUS X OBTUSIFOLIUS.
Waste places. Monongalia: streets of Morgantown.

R. SANGUINEUS, L. Bloody Dock.
Damp places in waste grounds. Berkeley: near Martinsburg.

R. CONGLOMERATUS, Murray.
Shady places. Fayette: near Nuttallburg—L. W. N.

R. ACETOSELLA, L. Horse Sorrel. J. L. W. N., V. M.. M. & G.
Abundant everywhere; even along paths in the dense spruce forests of the higher mountains.

ARISTOLOCHIACEÆ.

ASARUM, L.

A. Canadense, L.. Wild Ginger. "Colic Root." M. & G.
Rich woods. Wirt: near Burning Springs. Gilmer:

near Glenville—V. M.: Prof. Brown. Jefferson: near Flow inn Spring Mill. Fayette: near Nuttallburg—L. W. N Monongalia: near Uffington and Morgantown. McDowell: near Elkhorn. Mercer: Bluestone Je., and common throughout the State.

A. Virginicum, L.

Rich soil. Grant: near Bayard. Tucker: along Black Water. Gilmer: near Glenville—V. M.: Prof. Brown. Greenbrier: near White Sulphur Springs. McDowell: near Elkhorn. Mercer: Bluestone Je. and Bluefield.

ARISTOLOCHIA, L.

A. Serpentaria, L. Virginia Snakeroot. M. & G.

Rich woods. Wirt: near Burning Springs. Randolph: near Ford's and on Point Mountain. Gilmer: near Glenville- Prof. Brown. Grant: near Bayard. Tucker: near Davis. Mineral: near Keyser—W. Fayette: near Nuttallburg— L. W. N.

A. Sipho, L'Her. Dutchman's Pipe.

Rich woods. frequent throughout the State. Abundant in the following localities. Randolph: on Point Mountain. Grant: near Bayard. Gilmer: near Glenville -V. M., Prof. Brown. Fayette: near Nuttallburg—L. W. N. Mercer: Ingleside.

PIPERACEÆ.

SAURURUS, L.

S. cernuus. L. Lizard's Tail.

Streams. Jefferson: near Harper's Ferry—M. & G. Brooke: near Wellsburg. Fayette: near Nuttallburg—L. W. N.

LAURINEÆ.

SASSAFRAS, Nees.

S. officinale. Nees. Sassafras. G., L. W. N.. V. M.. M. & G.

Thickets and opens. Abundant throughout the State.

LINDERA, Thumb.

L. Benzoin(L.), Meisn. Wild Allspice. Spice-bush. L. W. N., V. M., M. & G.

Low woods. Common throughout the State.

THYMELÆACEÆ

DIRCA, L.

D palustris. L. Leatherwood.
Damp woods. Jackson: near Ripley. Wirt near Elizabeth. Calhoun: near White Pine and Brookville. Greenbrier: White Sulphur Springs. Fayette: near Nuttallburg—L. W. N.

LORANTHACEÆ.

PHORADENDRON, Nutt.

P. flavescens (Pursh.), Nutt. American Mistletoe.
Parasitic on Sugar-Maple and Black Locust, along the Great Kanawha River in Fayette county. On Black Walnut and Elm in Kanawha county. On Elms in Mason county, and in Wood near Parkersburg. On Elms and Hickories, along the Ohio and Great Kanawha rivers, in Cabell county.

SANTALACEÆ.

PYRULARIA, Michx.

P. pubera, Michx.
Rich woods. McDowell: near Elkhorn.

EUPHORBIACEÆ.

EUPHORBIA, L.

E. glytosperma Englm. *var.* **pubescens,** Englm.
Sandy soil. Mason: banks of the Ohio river, near Point Pleasant. The only station so far known in the State.

E. maculata. L. Spotted Spurge.　　　　L. W. N., M. & G.
Arid soils. Common even in the higher Alleghanies.

E. Preslii. Guss.　　　　　　　　　L. W. N., M. & G.
Dry soils and pastures. Common throughout the State, even in the higher Alleghanies.

E MARGINATA, Pursh.
An escape from cultivation. Taylor: permanently established near Mannington—V. M. Monongalia: the Flats near Morgantown.

E. corollata. L. Flowering Spurge. M. & G.
Dry soils. Wirt: near Elizabeth. Lewis: along
Leading Creek. Upshur: near Laurentz. Randolph: on
Lone Sugar Knob. alt. 2,800 ft. Webster: Long Glade.
Nicholas: Mumble-the-Peg Creek. Kanawha: along 8-Mile
Creek; near Pocataligo. Jackson: Fisher's Point. Wood:
on Limestone Ridge. Gilmer: near Glenville—V. M. Fay-
ette: near Nuttallburg. L. W. N. Grant: near Bayard.
Monongalia. plentiful along Cheat River. near Camp Eden.
Greenbrier: near White Sulphur Springs. Monroe. near
Alderson. Summers: near Hinton. Mason: near Point
Pleasant.

E. dentata. Michx.
Rich soil. Ohio: near Wheeling—M. & G.

E. Darlingtonii, Gray.
Damp woods. Pocahontas: along the mountains—
A. D. Hopkins.

E. obtusata. Pursh.
Rich soil. Ohio: near Wheeling—M. & G.

E. CYPARISSIAS. L. "Grave-yard-weed."
A frequent escape from cemeteries. Monongalia: near
Morgantown. Cabell: near Huntington.

E. commutata, Englm.
Woodlands. Jefferson: near Harper's Ferry—M. &
G.: near Shepherdstown. Mineral: near Keyser—W. along
Knobby Mountains. Summers: near Hinton. abundant.
Hardy: near Moorefield—G. Hampshire: near Romney.

E. LATHYRIS. L. "Mole-weed."
Escaped from gardens, where it is cultivated with the
idea of keeping out moles. Randolph: roadside up Point
Mountain.

ACALYPHA, L

A. Virginica, L. M. & G.
Fields and waste places. Common throughout the
state.

forma **intermedia**. mihi.
A form apparently uniting A. Virginica, L. with A.
Caroliniana. Ell. especially in the matter of leaves and bracts.
is found near Nuttallburg—L. W. N.. and Hawk's Nest. as
well as along New River opposite Hinton.

437

URTICACEÆ.

ULMUS. L.

U. fulva, Michx. Slippery Elm. M. & G.
Rich soils. Monongalia: near Morgantown, Laurel Point and Stumptown. Gilmer: near Glenville—V.M. Fayette: near Nuttallburg—L. W. N. Mason: near Point Pleasant. Summers: along Greenbrier river.

U. Americana, L. White Elm. L. W. N., V. M., M. & G.
Along rivers. Frequent throughout the State.

U. racemosa, Thomas. Corky Elm. M. & G.
Near streams. Monroe: near Alderson. Summers: along Greenbrier river.

CELTIS, L.

C. occidentalis, L. Hackberry.
Woods and river banks. Jefferson: near Shenandoah Junction. Monongalia: near Morgantown. Jackson: near Ripley. Fayette: near Nuttallburg—L. W. N.

CANNABIS, L.

C. SATIVA, L. Hemp.
Fields and waste places. Escaped from cultivation. Jackson: frequent.

HUMULUS, L

H. LUPULUS, L. Hops.
Alluvial banks near streams. Very doubtfully native. Randolph: near Cricard. Marshall: near Moundsville. Marion: near Clements, and Catawba. Mineral: opposite Cumberland, Md.

MORUS, L

M. rubra, L. Black Mulberry.
Rich woods. Wood, Wirt, Calhoun and Gilmer—V.M. along the Little Kanawha River. Jefferson: frequent throughout. Greenbrier: near White Sulphur Springs. Fayette: near Nuttallburg, no large trees noted—L. W. N.

M. ALBA, L. White Mulberry.
A frequent escape. Monongalia: near Morgantown. Jefferson: near Millville and Charlestown.

438

BROUSSONETIA.

B. PAPYRIFERA. Paper Mulberry. "Cut Paper."
Escaped from cultivation. Jefferson: near Flowing
Spring Mill and Milltown. Kanawha: near Montgomery.
Berkeley: near Martinsburg.

URTICA, L.

U. gracilis, Ait. Nettle.   V. M., M. & G.
Moist shady places. Common.

U. URENS, L. Stinging Nettle.
Adventive. Hancock: near Holliday's Cove. Rare.

LAPORTEA, Gaud.

L. Canadensis(L.),Gaud. Wood Nettle.
Moist rich woods. Fayette: near Nuttallburg—L. W.
N.: near Kanawha Falls—James. Frequent throughout the
State.

PILEA, Lindl.

P. pumila(L.), Gray. Clear Weed. Rich Weed. L.W.N., M.&G.
Cool, moist, shady places. Common throughout the
State.

BŒHMERIA, Jacq.

B. cylindrica(L.), Willd.
Damp places. Fayette: near Nuttallburg—L. W. N.
Monongalia: near Uffington.

PLATANACEÆ.

PLATANUS, L.

P. occidentalis, L. Sycamore. Buttonwood. L.W.N. V.M., M.&G.
All soils. Common throughout the State.

JUGLANDEÆ.

JUGLANS. L.

J. cinerea, L. Butternut. White Walnut. L.W.N., V.M., M.&G.
Common throughout the State.

J. nigra. L.   Black Walnut.   L. W. N., V. M., M. & G.
Rich soils, even in the higher Alleghanies. A very common and valuable timber tree throughout the State.

HICORIA, Raf.

H. ovata.(Mill.), Britt.   Shag or Shellbark Hickory (*Carya alba*, Nutt.)   L. W. N., V. M., M. & G.
Low grounds, frequent throughout the State.

H. sulcata,(Nutt.)   King Nut. (*Carya sulcata*, Nutt.)
Rich soil. Monongalia: near lee's Ferry.

H. alba.(L.)   Britt.   White Heart Hickory. (*Carya tomentosa*, Nutt.)   L. W. N., V. M.
Woods, frequent throughout the State.

H. glabra(Mill.), Britt.   Pig Nut. (*Carya porcina*, Nutt.)   L. W. N., V. M., M. & G.
Dry soils, frequent throughout the State.

H. microcarpa(Nutt.), Britt.   (*Carya microcarpa*, Nutt.)
Woodlands.   Fayette: near Nuttallburg—L. W. N.

H. minima(Marsh.), Britt.   Bitter-nut.   *Carya amara*, Nutt.
Low Woods. Greenbrier: near Fort Spring and Ronceverte.

CUPULIFERÆ.

BETULA, L.

B. lenta, L.   Sweet Birch.   Black Birch.   L. W. N., M. & G.
Rich Woods. Common throughout the State. Grows very large in the mountains. One specimen near Cheat Bridge measure: 7 ft. 9 in. in diameter.

B. lutea, Michx.f.   Yellow Birch.
Higher mountain woods.   Grant: near Bayard. Tucker: near Hulings. Braxton: near Sutton. Randolph: near Pickens.

B. populifolia. Marsh.   White Birch.
Poor soils. Gilmer: near Glenville V. M. Randolph: near Winchester.

B. nigra, L.   River Birch.   Red Birch.
Along streams. Calhoun: along Little Kanawha River. Gilmer: near Glenville—V. M. Greenbrier: near

Fort Spring. Summers: near Greenbrier Stockyards: near
Hinton. Kanawha: near Handley. Fayette: near Nuttallburg—L. W. N.; and common along streams throughout the
central and southern counties.

ALNUS, L.

A. viridis, DC. Mountain Alder.
Along mountain streams, rare. Greenbrier: Colum-
bia Sulphur Springs. Fayette: near Nuttallburg—L. W. N.
Pocahontas: at Traveler's Repose. Randolph: along Cheat
River.

A. serrulata, Willd. Smooth Alder.
Low grounds and along rivers. Common, especially
in the glade regions.

CARPINUS, L.

C. Caroliniana, Walt. Blue or Water Beech.
Damp thickets and river banks. Wirt: near Elizabeth.
Monongalia: near Morgantown, plentiful. Gilmer: near
Glenville—V. M. Fayette: near Nuttallburg—L. W. N.
Summers: near Hinton. Marion: near Worthington. Jef-
ferson: near Harper's Ferry—M. & G.

OSTRYA, Scop.

O. Virginiana(Mill.), Willd. Lever Wood. Iron Wood. M. & G.
Rich woods and along streams. Wirt: near Elizabeth.
Randolph: on Point Mountain: at first Top of Cheat there
is a forest of this wood where trees are found in quantity
from 1–3 feet in diameter. Webster: on Buffalo Bull
Mountain. Greenbrier: near White Sulphur Springs. Summers: near Hinton. Marion: near Worthington. Fayette:
near Nuttallburg—L. W. N.

CORYLUS. L.

C. Americana, Walt. Hazlenut.   V. M.. M. & G.
Thickets, frequent throughout the State.

C. rostrata. Ait. Beaked Hazlenut.
Mountainous regions. Upshur: summit on Staunton
Pike. Randolph: near Fords.

QUERCUS, L.

Q. alba. L.. White Oak.   L. W. N.. V. M.. M. & G.
All soils, plentiful throughout the State.

Q. minor (Marsh.), Sarg.   Post Oak.   Iron Oak.   (*Q. stellata*, Wang.)
Dry sterile soils, common.

Q. macrocarpa, Michx.   Burr Oak.   Mossy-cup Oak.
Rich soils.   Tyler : near Long Reach.

Q. Prinus. L.   Chestnut Oak.   L. W. N., M. & G.
Rocky woods.   Frequent throughout the State.

Q. Muhlenbergii, Engelm.   Yellow Oak.
Rich, wooded valleys, especially in the mountains.
Fayette : near Nuttallburg. rare—L. W. N.

Q. rubra, L.   Red Oak.   L. W. N., V. M., M. & G.
Common throughout the State, in both rich and poor
soils.

Q. coccinea, Wang.
Woodlands.   Fayette : near Nuttallburg, apparently a
second growth—L. W. N.

Q. tinctoria, Bartr.   Black Oak.
Dry woodlands.   In large tracts in the Alleghanies of
Mineral, Grant and Tucker counties.   Gilmer : near Glenville —V. M.   Fayette : near Nuttallburg—L. W. N.   Monongalia : near Ice's Ferry.   Summers : near Hinton.

Q. palustris, DuRoi.
Along streams.   Monongalia : near Stumptown.

Q. cuneata, Wang.   Spanish Oak.   *Q. falcata*. Michx.
Dry sandy soils throughout the western counties.

Q. nigra, L.   Black Jack Oak.   V. M.
Common in dry or heavy clay soils throughout the
center of the State.   Hardy : near Moorefield.

Q. ilicifolia, Wang.   Holly-leaved Oak.
Sandy soils.   Hampshire : near Romney.   Hardy :
near Moorefield.

Q. imbricaria, Michx.   Laurel Oak.
Rich woods.   Monongalia : near Morgantown and
Laurel Point.   Hardy : near Moorefield.

CASTANEA, Gaertn.

C. sativa, Mill., *var.* **Americana** (Michx.), Sargent.   Chestnut.
                                                          L. W. N.
Rocky woods and hills throughout the State.

C. pumila, Mill. Chinquapin.
Dry hills. Fayette: near Nuttallburg, alt. 2000 ft.,
frequent—L. W. N. Wayne: near Ceredo and Compton's
Creek. Mercer: Beaver Spr., and Ingleside.

FAGUS, L

F. ferruginea. Ait. Beech. L. W. N., V. M., M. & G.
General.

SALICINEÆ.

SALIX, L.

S. nigra, Marsh. Black Willow. L. W. N., M. & G.
Along streams, frequent or common. The principal
willow of the State.

Var. **falcata**. Torr. Scythe-leaved Black Willow.
Along springy runs. Wirt: along Straight Creek.
Lewis: along Leading Creek. Fayette: near Nuttallburg—
L. W. N.

S. amygdaloides, And.
Fayette: near Nuttallburg—L. W. N.

S. ALBA, *var.* **VITELLINA**. Koch. White Willow. M. & G.
Scattered along streams in many parts of the State.

from protective or ornamental planting.

S Babylonica, Tourn. Weeping Willow.
A frequent escape as in the last species. Monongalia:
near Morgantown. Jefferson: near Flowing Spring and
Milltown.

S. discolor, Muhl. Shining Willow.
Ohio: on Bogg's Island—M. & G.

S humilis, Marsh. Prairie Willow.
Glady regions. Webster: near Upper Glade. Preston:
near Terra Alta.

S. sericea. Marsh. Silky Willow.
Along streams. Randolph: along Tygart's Valley
River. Greenbrier: near White Sulphur Springs.

S. cordata, Muhl. Heart-leaved Willow.
Along streams, frequent. Lewis: along Leading

443

Creek. Wood: near Parkersburg. Mason: near Point Pleasant.

POPULUS. L.

P. ALBA. L. White Poplar. Abele.　　　　　　V. M.
　　A frequent escape from cultivation. In many places in the State, where the tree is planted for ornament, this species spreads widely from the root, thus often becoming a pest in lawns and along streets.

P. tremuloides, Michx. Aspen. Trembling Poplar. M. & G.
　　Wooded hillsides. Wirt: along Little Kanawha River. Calhoun: near Grantsville. Gilmer: near Glenville. Monongalia: near Marion: along the Monongahela River. Summers: near Riffe. Mason: near Point Pleasant:

P. grandidentata, Michx. Large-toothed Aspen.
　　Preston: near Cranberry Summit—M. & G. Ohio: near Wheeling—M. & G.

P. balsamifera. L., *var.* **CANDICANS**(Ait.). Gray. Balm of Gilead. M. & G.
　　Plentiful at Montana, along the Monongahela river in Marion Co. Monongalia: the Flats near Morgantown. Gilmer: near Glenville—V. M. Fayette: near Nuttallburg, very likely introduced—L. W. N.

P. monilifera, Ait. Cottonwood.
　　Ohio: near Bogg's Run—M. & G. Mason: near Point Pleasant.

COREMA CONRADI. Torr., mentioned in Botanical Gazette. Vol. 2., P. 136, as occuring near Hawk's Nest, Fayette Co., is proven by Prof. James to be another species: and should therefore not yet be credited to this State.

———O———

MONOCOTYLEDONÆ.

HYDROCHARIDEÆ.

ELODEA, Michx.

E. Canadensis, Michx. Water-weed.
Slow streams in slack water. Fayette: near Kana
wha Falls—James. Putnam: near Buffalo. .

ORCHIDEÆ.

MALAXIS, Sw. (1788).
(Microstylis, Nutt. 1818).

M. unifolia Raf.) M. *ophioglossoides*, (Nutt.
Rich woodlands. Greenbrier: near White Sulphu
Springs—M. & G. Fayette: near Nuttallburg, rare—L.W.N

LEPTORCHIS, Thou. (1809).
(Liparis, Rich. 1818).

L. liliifolia(L.), Tway-blade.
Rich woods. Monongalia: the Flats near Morgan
town. Gilmer: near Glenville- V. M., Prof. Brown.

APLECTRUM, Nutt.

A. spicatum(Walt.). B. S. P. Adam-and-Eve. (*A. hiemale*, Nutt.
Rich woods. Monongalia: rich woods near Morgan
town. Gilmer: near Glenville—V. M. Fayette: near Nut
tallburg- L. W. N.

CORALLORHIZA, R. Br.

C. innata, R. Br. Coral Root.
Deep, rich woods. Grant: near Bayard. Fayette
near Nuttallburg, alt. 2000 ft,— L. W. N. Greenbrier: nea
White Sulphur Springs.

C odontorhiza, Sw., Nutt.
Rich woods. Gilmer: near Glenville—V. M. Fay
ette: near Nuttallburg, alt. 2000 ft. —L. W. N.

C. multiflora, Nutt
Woodlands—Fayette: near Nuttallburg, alt. 2000 ft,
L. W. N.

GYROSTACHYS, Pers. (1807.)
(Spiranthes. Rich. 1818)

G. cernua(L.) Ladies-Tresses.
Wet meadows. Fayette: near Nuttallburg. L. W. N.
Monongalia: frequent throughout.

G. gracilis(Bigel.) Twisted Stock.
Sandy woods and fields. Fayette: near Nuttallburg,
in hard, grassy ground—L. W. N. Monongalia: near Morgantown. Frequent throughout the State.

GOODYERA, R. Br.

G. repens, R. Br.
In deep evergeen forests. Grant: under Black Spruce
near Bayard. Fayette: in deep Laurel thickets, rare, near
Nuttallburg—L. W. N. McDowell: near Elkhorn.

G. pubescens(Willd.) Rattlesnake Plantain. M. & G.
Rich woods. Monongalia: along Decker's Creek and
Tibb's Run. Fayette: near Nuttallburg, rare,-L. W. N.
Nicholas: near Beaver Mills. McDowell: near Welch.

CALOPGON, R. Br.

C. tuberosus(L.). B. S. P. *C. pulchellus,* R. Br.
Hardy: near Moorefield—G.

POGONIA, Juss.

P. ophioglossoides (L.), Ker.
Boggy places. Preston: near Cranberry Summit—
M. & G.

ORCHIS, L.

O. spectabilis, L. Showy Orchis.
Rich woods. Monongalia: near Morgantown. Gil-
mer: near Glenville—N. M., Prof. Brown. Fayette: near
Nuttallburg, rare—L. W. N. McDowell: near Elkhorn.

HABENARIA, Willd.

H. tridentata (Willd.). Hook. Rein Orchis.
Wet places. Webster: in Long Glade. Fayette:
near Nuttallburg—L. W. N.

H. flava (L.). Gray.   *H. virescens.* Spr.
Wet places. Webster: in Long Glade.

H. orbiculata (Pursh). Torr.
Rich woods. Randolph: near summit Rich Mountain.

H. ciliaris (L.). R.Br. Yellow Fringed Orchis.
Wet sandy bogs. Webster: Welch and Long Glades.
Preston: near Terra Alta. Fayette: near Nuttallburg, alt.
2000 ft., rare—L. W. N. Hardy: near Moorefield—G.

H. lacera (Michx.). R.Br. Ragged Fringed Orchis.
Bogs and moist thickets. Wood: near Lockhart's
Run. Preston: Cranberry Summit—M. & G.

H. psycodes (L.). Gray. Pink Fringed Orchis.
Wet places. Randolph: on Rich Mountain. Grant:
near Bayard. Preston: near Terra Alta. Wayne: near
Central City.

CYPRIPEDIUM. L.

C. parviflorum, Salisb. Small Lady's Slipper.
Rich woods. Monongalia: near Morgantown. Marion: near Opekiska and Fairmont. Gilmer: near Glenville
—V. M.: Prof. Brown. Mineral: near Keyser—W. Hardy:
near Moorefield—G.

C. pubescens, Willd. Large Lady's Slipper.
Low woods. Same stations as the last. Fayette:
near Nuttallburg—L. W. N.

C. acaule. Ait. Moccasin Flower.
Dry or moist woods. Monongalia: near Cheat View,
plentiful. Preston: near Reedsville. Gilmer: near Glenville—Prof. Brown. Fayette. near Nuttallburg, alt. 2000 ft.,
in laurel thickets—L. W. N. Marion: along the F. M. &
P. R. R. Mineral: near Keyser—W. Kanawha: near Coalburg James. Hardy: near Moorefield—G.

HÆMODORACEÆ.

ALETRIS, L.

A farinosa, L. Star Grass. Colic Root.
Sandy moist soils. Webster: in Upper Glade. Fayette: near Nuttallburg—L. W. N.

IRIDEÆ.

IRIS, L.

I. versicolor, L. Blue Flag.
Ditches and wet lands. Preston: near Terra Alta and Reedsville.

I. verna, L. Dwarf Iris.
Hardy: near Moorefield— G.

I. cristata. Ait. Dwarf Crested Iris.
Rich woods. Monongalia: near Morgantown, plentiful. Gilmer: near Glenville—V. M., Prof. Brown. Fayette: near Nuttallburg L. W. N. Marion: along the Monongahela River, plentiful.

SISYRINCHIUM, L.

S. angustifolium. Mill. Blue-eyed Grass. *S. Bermudianum*, Gray. not. L. M. & G.
Moist grassy places. Fayette: near Kanawha Falls— James: near Nuttallburg—L. W. N. Gilmer: near Glenville—V.M. Monongalia: near Morgantown, and elsewhere common.

S. anceps, Cav.
Grassy places. Fayette: near Nuttallburg—L. W. N.

AMARYLLIDEÆ.

HYPOXYS. L.

H. erecta, L. Star-Grass. Yellow-eyed Grass.
Open places. Fayette: near Nuttallburg, common— L. W. N. Lewis: Up Stone Coal Creek. Monongalia: along Cheat River near Camp Eden, plentiful. Hardy: near Moorefield— G. Mercer: near Bluefield.

DIOSCOREACEÆ.

DIOSCOREA. L.

D. villosa, L. Wild Yam-root. Colic-root. M & G.
Thickets and rich woods. Randolph: on Point Mountain, alt. 3450 ft. Gilmer: near Glenville—V. M. Fayette: near Nuttallburg. Pistillate plants rare—L. W. N. Grant and Tucker plentiful in the mountain woods. Mon-

ongalia: plentiful. Greenbrier: near White Sulphur
Springs. Summers: near Hinton. Hardy: near Moorefield.
McDowell: near Elkhorn. Mercer: Bluestone—J. C.

LILIACEÆ.

SMILAX, L.

S. herbacea, L. Carrion Flower. M. & G.
Thickets. Monongalia: plentiful at the Flats. Fayette: near Nuttallburg, rare—L. W. X.

S. rotundifolia, L. Greenbrier. L. W. X., V. M., M. & G.
Thickets, too common throughout the State.

S. glauca, Walt.
Dry thickets. Kanawha: near Charleston—Barnes.
Fayette: near Nuttallburg—L. W. X. Monongalia: along Cheat River near Ice's Ferry.

S. Pseudo-China, L.
Dry soils. Monongalia: plentiful on the Flats; near Morgantown.

S. hispida, Muhl. Bristly Sarsaparilla.
Rich soil. Marion: near Fairmont.

ASPARAGUS, L.

A. OFFICINALIS, L. Asparagus. M. & G.
Woods and opens: a frequent escape. Jefferson: near Shenandoah Junction, plentiful. Monongalia: plentiful along Cheat River, near Ice's Ferry, and at Stewartown.
Mason: near Point Pleasant. Berkeley: near Martinsburg.
Hardy: near Moorefield.

POLYGONATUM, Adans.

P. biflorum(Walt.), Ell. Small Solomon's Seal. G., L. W. X.,
M. & G.
Rich woods and wooded banks. Frequent generally.

P. commutatum(Schult.), Dietr. Solomon's Seal. (*P. giganteum*
Dietr.)
Edges of meadows. Nicholas: near Beaver Mills.
Gilmer: near Glenville—V. M. Monongalia: the Flats near Morgantown. Fayette: near Nuttallburg—L. W. X.

STREPTOPUS. Michx.

S. roseus, Michx. Twisted Stalk.
Cold, damp woods. Upshur: along Sand Run. Preston: near Terra Alta. Tucker: near Blackwater Falls. Grant: near Bayard. Randolph: western slopes of Cheat Mountains.

DISPORUM. Salisb.

D. lanuginosum, Benth. & Hook. *Prosartes lanuginosa,* Don.)
Rich woods. Monongalia: near Morgantown. Preston: near Terra Alta. Mineral: near Keyser –W. Fayette: near Nuttallburg—L. W. N. Grant: near Bayard. Mercer: Bluestone Je.

UNIFOLIUM, Adans. (1763)
(Smilacina, Desf. 1798.)

U. racemosum(L.), Britt. False Solomon's Seal. L.W.N., V.M., M. & G.
Rich woods. Common throughout the State.

U. Canadense(Desf.), Greene. Wild Lily-of-the-Valley (*S. bifolia,* var. *Canadensis,* Gray). M. & G.
Low, rich woods. Common throughout the northern and eastern counties.

MAIANTHEMUM. Wigg.

M. Canadense. Desf.   M. & G.
Rich woods. Fayette: near Nuttallburg—L. W. N. Monongalia: along Decker's Creek.

HEMEROCALLIS. L.

H. FULVA, L. Day-lily. "Eve's Thread".
Roadsides and fields. A frequent escape from cultivation. Monongalia: near Morgantown and Cassville. Hampshire: near Bloomery where "it has become widely scattered by ploughing."

ALLIUM. L.

A. VINEALE. L. Field Garlic.
Cultivated fields. Jefferson: a vile and abundant weed in wheat fields. Monongalia: along the Monongahela below Morgantown.
Reported as a weed from: Berkeley: near Oakton, Martinsburg and Hedgesville. Barbour: near Belington,

Braxton: near Lloydsville. Cabell: near Union Ridge. Doddridge: near Smithton. Grant: near Medley. Greenbrier: near Trout Valley and White Sulphur Springs. Hampshire: near Slanesville. Three Churches. Capon Bridge, Bloomery. Dillon's Run. Higginsville and Springfield. Harrison: near Good Hope and Bridgeport. Hardy: near Wardensville and Mooresfield. Jefferson: near Summit Point, Mohler's. Leetown. Charlestown. Ripon. Middleway and Kabletown. Jackson: near Sandy. Kanawha: near Blandon. Lincoln: near Hamlin. Mercer: near Princeton, and Concord Church. Marion: near Barracksville and Gray's Flat. Monongalia: near Morgantown. Ohio: near West Liberty. Preston: near Terra Alta, Reedsville, Amblersburg and Independence. Ritchie: near Ritchie C. H. Roane: near Walnut Grove. Summers: Talcott. Taylor: near Knottsville. Thornton and Grafton. Upshur: near Kanawha Head. Wayne: near Adkin's Mills. Wetzell: near Endicott and Blake. Wirt: near Reedy Ripple. Wool: near Waverly. Blennerhassett and Rockport.

A. tricoccum, Ait. Wild Leek. "Ramps." M. & G.
Rich mountain woods. Grant: near Bayard, abundant. Tucker: abundant along Blackwater Fork of Cheat. Greenbrier: near White Sulphur Springs.

A. cernuum, Roth. Wild Onion.
Greenbrier: near White Sulphur Springs. Fayette: near Nuttallburg—L. W. N. Ohio: near Wheeling—M. & G. Summers: near Hinton. Monongalia: near Camp Eden, and banks of Decker's Creek.

A. Canadense. Kalm. Wild Garlic.
Ohio: near Wheeling—M. & G. Fayette: near Loup Creek—James.

CAMASSIA, Lindl.

C. Fraseri, Torr. Wild Hyacinth.
Rich ground. Ohio: near Wheeling—M. & G. Gilmer: near Glenville—V. M.

MUSCARI, Mill.

M. BOTRYOIDES (L.. Mill. Grape Hyacinth.
Escaped from gardens. Monongalia: near Uffington. Gilmer: near Glenville—V. M.

ORINTHOGALUM, L.

O. UMBELLATUM, L. Star of Bethlehem. M. & G.
Escaped from gardens. Monongalia: abundant and

451

persistent on the line of the F., M. & P. R. R. from Morgantown to Coburn's Creek. Gilmer: near Glenville—V. M.

LILIUM, L.

L. Philadelphicum, L. Wild Red Lily. "Glade Lily." V. M.,
M. & G.
Dry or damp grounds. Monongalia: in Clinton District M. H. Brown; near Stewartown, and Lee's Ferry. Marion: along the F., M. & P. R. R. Gilmer: near Glenville—V. M. Webster: in the glades, where it is called "Glade Lily." Hardy: near Moorefield—G. Fayette: near Nuttallburg—L. W. N.

L. superbum, L. Turk's-cap Lilly. M. & G.
Rich, low grounds. Randolph: summit Point Mountain, alt. 3700 ft. Monongalia: near Morgantown and Stewartown. Hardy: near Moorefield—G.

L. Canadense, L. Wild Yellow Lily. M. & G.
Moist meadows. Calhoun: along Laurel Run. Gilmer: near Glenville—V. M. Randolph: summit Point Mountain, alt. 2125 ft. Monongalia: near Lee's Ferry.

L. TIGRINUM, Ker. Tiger Lily. M. & G.
Established from gardens. Jefferson, Berkeley, Morgan, Hampshire and Mineral: along the B. & O. R. R. Calhoun: near Brookville. Monongalia: near Morgantown.

ERYTHRONIUM, L.

E. Americanum, Ker. Yellow Adder's Tongue. Dog's Tooth Violet. M. & G.
Rich open woods along streams. Monongalia and Marion: along the Monongahela River. Gilmer: near Glenville—V. M., Prof. Brown. Fayette: near Nuttallburg—L. W. N. Frequent or common throughout the State. Hardy: near Moorefield—G.

CHAMÆLIRIUM, Willd.

C. luteum (L.), Gray. Blazing Star. Devil's-bit.
Glades. Preston: near Terra Alta and Kingwood. Webster: in Welsh, Long and Collett's Glades. Gilmer: near Glenville—V. M. Fayette: near Nuttallburg L. W. N.

UVULARIA, L.

U. perfoliata, L. Bellwort.
Rich woods. Marion and Monongalia: along the

Monongahela river. Gilmer: near Glenville—V. M.. Prof.
Brown. Fayette: near Nuttallburg—L. W. N.

U. sessilifolia. L. *Oakesia sessilifolia* (L.), Wats.
Low rich woods. Gilmer: near Glenville—V. M.
Fayette: near Nuttallburg—L. W. N. Mercer: near Beaver
Spr.

CLINTONIA. Raf.

C. borealis(Ait.), Raf. Clinton's Lily.
Cold damp woods. Grant: near Bayard. Preston:
in Laurel hills. Tucker: near Davis.

C. umbellata, Torr.   M. & G.
Rich woods. Randolph: on Rich Mountains. Grant:
near Bayard Tucker: in Land of Canaan. Gilmer: near
Glenville—Prof. Brown. V. M. Monongalia: near Morgan-
town. Kanawha: near Coalburg—James. Fayette: near
Nuttallburg—L. W. N. Hardy: near Moorefield—G. Mc-
Dowell: near Elkhorn. Mercer: Bluestone Jc.

MEDEOLA. L.

M. Virginica. L. Indian Cucumber.   M. & G.
Rich, damp woods. Upshur: along Sand Run. Grant
and Tucker: in the mountain woods, plentiful. Gilmer:
near Glenville—V. M. Monongalia: near Morgantown.
Kanawha: near Coalburg—James. Fayette: near Nuttallburg—L. W. N., and elsewhere frequent. McDowell: near
Elkhorn.

TRILLIUM, L.

T. sessile, L. Sessile-flowered Purple Trillium.   M. & G.
Moist woods. Monongalia and Marion: abundant
along the Monongahela river. Gilmer: near Glenville—Prof.
Brown.

T. erectum. L. Purple Trillium.   L. W. N., V. M., M. & G.
Rich woods. Throughout the northern and eastern
portions of the State. McDowell: near Elkhorn.

Var. **album.** Pursh.
Monongalia: near Morgantown. Gilmer: near Glen-
ville— V. M., Prof. Brown. Taylor: near Valley Falls—N.
S. Hayes.
(See remarks under next form.)

Var. **declinatum.**
Gilmer: near Glenville—V. M. Mason near Point
Pleasant. Jackson: near Ravenswood.
(NOTE: This form appears to be sufficiently different
from T. erectum, and T. cernuum, to retain it; otherwise I
could not list the specimens gathered.)

T. grandiflorum, Salisb. White Trillium. M. & G.
Rich woods. Abundant throughout the northern
counties. Gilmer:near Glenville—V. M., Prof. Brown. Fayette: near Nuttallburg—L. W. N. McDowell: near Elkhorn.

T. cernuum. L. Wake Robin.
Moist woods. Gilmer: near Glenville—V. M. Fayette: near Nuttallburg—L. W. N.

T. nivale. Riddell. Dwarf White Trillium.
In moss of wet rocks. Monongalia: along Quarry
Run. Grant: along Buffalo Creek. Randolph: along runs
feeding Shaver's Fork of Cheat. Tucker: along Beaver
Creek, and Blackwater Fork of Cheat.

T. erythrocarpum, Michx. Painted Trillium.
Cold deep ravines along runs. Grant: near Bayard.
Randolph: Cheat River: Tucker: along Black Water Fork.
Fayette: near Nuttallburg. alt. 2000 ft., rare—L. W. N.
McDowell: near Elkhorn. Taylor: near Valley Falls—N.
S. Hayes. Mercer: Bluestone Je.

MELANTHIUM. L.

M. Virginicum, Bunch-flower.
Glades. Preston: near Terra Alta and in Morgan's
Glade. Monongalia: glades near Booth's Creek. Webster:
Second Glade. Nicholas: Collett's Glade.

M. parviflorum(Michx.). Watson.
Greenbrier: near White Sulphur Springs—M. & G.

VERATRUM, L.

V. viride. Ait. American White Hellebore. M. & G.
Rich wet spots in deep mountain woods. Randolph:
on Point Mountain. Webster: Buffalo Bull Mountain. Grant
and Tucker: common along streams.

CHROSPERMA. Raf. (1825)
(Amianthium, Gray 1837)

C. muscætoxicum (Walt.) Fly Poison.
Low rich grounds. Monongalia: Ice's Ferry. Preston: near Terra Alta.

COMMELINACEÆ.

TRADESCANTIA. L.

T. Virginica, L. Spiderwort.
Rich grounds. Kanawha: near Charleston—Barnes.
Gilmer: near Glenville—V. M. Fayette: near Nuttallburg
—L.W.N. Wirt, along Straight Creek and near Burning
Springs. Hardy: near Moorefield—G.

T. pilosa. Lehm. Hairy Spiderwort.
On boulders. Wirt: abundant beyond Burning
Springs, and along Straight Creek.

COMMELINA, L.

C. Virginica, L. Day Flower.
Damp opens. Fayette: near Nuttallburg, rare—L.W.N.

JUNCACEÆ.

JUNCUS, L.

J. effusus, L. Soft Rush.   M. & G.
Marshy ground. Kanawha: near Charleston—James.
Fayette: near Nuttallburg—L. W. N. Common throughout.

J. setaceus, Rostk.
Fayette: near Loup Creek(?)—James.

J. marginatus, Rostk.
Moist sandy soils. Kanawha: near Charleston—
James. Common throughout.

Var. **paucicapitatus,** Englm.
Fayette: near Nuttallburg—L. W. N.

J. tenuis, Willd. "Poverty Grass."   L. W. N.
Roadsides and ditches. Abundant throughout the
State.

J. acuminatus, Michx　　　　　　　　L. W. N., M. & G.
　　Damp places. Monongalia and Marion; on the Monongahela River. Webster: in Welch and Long Glades, Fayette: near Nuttallburg.

J. nodosus, L.
　　Fayette: near Nutta o ᵉ—L. W. N.

J. Canadensis, J. Gay.
　　Monongalia: near Little Falls. Fayette: near Nuttallburg—L. W. N.

LUZULA. DC.

L. vernalis L.). DC.
　　Ohio: on Thomas's Hill near Wheeling—M. & G. Fayette: near Nuttallburg—L. W. N.

L. campestris L.). DC.
　　Ohio: near Benwood—M. & G. Tyler: near Long Reach. Fayette: near Nuttallburg—L. W. N. Monongalia: near Morgantown.

TYPHACEÆ.

TYPHA, L.

T. latifolia, L. Cat Tails.
　　Ditches and swampy spots. Gilmer: near Glenville—V. M. Preston: near Terra Alta. Morgan: near Hancock. Monongalia: near the Dille farm.

SPARGANIUM, L.

S eurycarpum, Engelm. Bur-reed.
　　Borders of streams and still waters. Webster: in Welch Glade. Greenbrier: near White Sulphur Springs.

AROIDEÆ.

ARISÆMA. Mart.

A. triphyllum (L.). Torr. Jack-in-the-Pulpit.　V. M., L. W. N., M. & G.
　　Rich, damp woods. Common throughout the State.

A. Dracontium (L.). Schott. Green Dragon.
　　Low grounds. Gilmer: near Glenville—V. M.

SPATHYEMA. Raf. (1808.)
(Symplocarpus. Salisb. 1812.)

S. foetida(L.). Raf. Skunk Cabbage.
Boggy meadows. Preston: near Terra Alta. ' Monongalia: on the Dille farm.
Reported from: Brooke: near Wellburg and Fowlers. Berkeley: near Martinsburg and Hedgesville. Barbour: near Belington. Cabell: near Milton. Grant: near Medley. Greenland, Mount Storm, Maysville and Petersburg. Greenbrier: near Trout Valley and Fort Spring. Hampshire: near Slanesville and Capon Bridge. Harrison: near Lost Creek, Bridgeport and Good Hope. Hardy: near Wardensville and Old Fields. Hancock: near Fairview and New Cumberland. Jefferson: near Kabletown and Summit Point. Jackson: near Odaville. Lewis: near Walkersville. Mercer: near New Bramwell and Princeton. Mason: near Maggie. Mineral: near Blaine. Marion: near Gray's Flats. Marshall: near St. Joseph and Glen Easton. Ohio: near West Liberty. Pocahontas: near Lobelia. Pleasants: near Schultz. Preston: near Terra Alta, Masontown, Eglon and Amblersburg. Randolph: near Lee Bell, Kerens and Cricard. Roane: near Clio. Summers: near Talcott and Clayton. Taylor: near Knottsville. Upshur: near Kanawha Haad and Hemlock. Wayne: near Adkin's Mills. Wetzel: near Littleton, Pine Grove and Blake. Wirt: near Burning Spring, Evelyn and Reedy Ripple. Wood: near Rockport. Webster: near Replete.

ACORUS, L,

A. Calamus L. Calamus. ' Sweet-flag.   M. & G.
Swampy places. Lewis: along Stone Coal Creek. Randolph: near Valley Bend, Cricard, and on Point Mountain, alt. 3050 ft. Gilmer: near Glenville—V. M. Berkeley: near Martinsburg.

LEMNACEÆ.

LEMNA. L.

L. minor, L. Duck-weed.
Stagnant waters. Putnam: near Buffalo.

ALISMACEÆ.

ALISMA. L.

A, Plantago. L. Water Plantain.   M. & G.
Wet ditches. Monongalia: near Ice's Ferry. Sum-

mers: near Hinton. Greenbrier: near White Sulphur
Springs.

SAGITTARIA. L.

S. sagittæfolia. L.. *forma* **hastata** Pursh) Britt.   Arrow-head.
                                                       M. & G.
   In water or very wet places.   Randolph: near Bev-
erly.   Summers: near Hinton.   Greenbrier: near White
Sulphur Springs.   Wood: near Parkersburg.   Marion: near
Clements.   Fayette: near Nuttallburg —L. W. N.   Berkeley:
near Martinsburg

forma **angustifolia**(Engelm.), Britt.
   Wet places.   Putnam: near Buffalo.   Mason: near
Point Pleasant.

S. graminea. Michx.
   Summers: shores of New River near Hinton.

N A I A D A C E Æ.

POTAMOGETON, L

P. fluitans. Roth.   Pond Weed.   *P. lonchites*, Tuck.
       In rivers and streams.   Randolph: near Tygart's Val-
ley River. near Hutton-ville.   Summers: New River: near
Hinton:

C Y P E R A C E Æ.

CYPERUS, L.

C. flavescens. L.   Galingale.
       Low grounds.   Monongalia and Marion: along the
Monongahela River.   Mason: near Point Pleasant.   Fayette: near Nuttallburg—L. W. N.

C. diandrus. Torr.
       Fayette: near Nuttallburg—L. W. N.

C. esculentus. L.   Edible Galingale.
       Preston: near Rowlesburg —M & G.

C. strigosus. L.   Lank Galingale.
       Low grounds.   Jackson: near Sandyville.   Mononga-
lia: near Morgantown.   Fayette: near Nuttallburg—L.W.N.

C. refractus, Engelm.
       Fayette: near Hawk's Nest—Porter.

C. Lancastriensis. Porter.
Summers: shores of New River, near Hinton.

KYLLINGA, Rott.

K. pumila, Michx.
Fayette: near Nuttallburg—L. W. N.

DULICHIUM, Pers.

D. spathaceum (L.). Pers.
Along streams. Randolph: Tygart's Valley River
near Huttonsville. Fayette: near Nuttallburg—L. W. N.

ELEOCHARIS, R. Br.

E. tuberculosa (Michx.). R. & S.
Sands of New River. Fayette: near Nuttallburg—L.
W. N.

E. ovata (Roth.), R. & S.   M. & G.
Muddy places. Upshur: near Buckhannon. Ran-
dolph: Tygart's Valley River. Fayette: near Kanawha
Falls—James: near Nuttallburg—L. W. N.

E. palustris (L.). R. & S.
Ohio: near Wheeling —M. & G. Fayette—near Nuttallburg—L. W. N.

Var. **glaucescens** (Willd.), Gray.
Jefferson: near Harper's Ferry—M. & G

E tenuis (Willd.), Schult. "Kill Cow." "Poverty Grass." M. &
                                                  G., L. W. N.
Damp places, abundant everywhere.
Note.— The above names are applied along Tygart's
Valley River, where in places this species grows so abundant
as to take whole fields, and as cattle will not thrive upon it,
these names suggested themselves.

E acicularis (L.). R. & S.
Jefferson : near Harper's Ferry—M. & G.

FIMBRISTYLIS, Vahl.

F. autumnalis (L.). Roem. & Schult.
Fayette: near Nuttallburg —L. W. N.

SCIRPUS. L.

S. pungens, Vahl.
Jefferson: near Harper's Ferry   M. & G.

S. lacustris. L.
Glady places.   Fayette: near Nuttallburg— L. W. N.

S. sylvaticus. L.
Glady places.   Fayette: near Nuttallburg —L. W. N.

S. atrovirens, Muhl.   M. & G.
Boggy places.   Monongalia: along the Monongahela River.   Fayette: near Nuttallburg — L. W. N.

S. polyphyllus. Vahl.   L. W. N.
Randolph: along Tygart's Valley River.   Fayette: near Nuttallburg.

ERIOPHORUM. L.

E. lineatum Michx.. Benth. & Hook.   Cotton Grass.
Low grounds.   Monongalia: along Falling Run, above the campus.

E. cyperinum L.   Wool Grass.   L. W. N.
Wet meadow lands.   Monongalia, frequent.   Upshur: near Buckhannon.   Webster: in Long Glade.   Kanawha: up 8-Mile Creek.

E. Virginicum. L.   L. W. N.
Damp places.   Fayette: near Nuttallburg.   Preston: near Cranberry Summit   M. & G.

RYNCHOSPORA, Vahl.

R. glomerata L.., Vahl.   Beak-rush.   L. W. N.
Low grounds.   Webster: Upper Glade.   Monongalia: near Morgantown.   Fayette: near Nuttallburg.

CAREX. L.

All the species of this genus have passed through the hands of Prof. Bailey, who kindly identified them for this Flora.

C. folliculata, L.
Margins of streams.   Tucker: near Falls of Blackwater.

C. intumescens. Rudge.
Wet places. Fayette: near Nuttallburg. Tucker: near Falls of Blackwater. Preston: near Terra Alta.

C. Grayii, Carey. Gray's Sedge.
Meadows and copses. Upshur: near Beech Fork.

C. lupulina. Muhl. Hop Sedge.
Wet places. Upshur: near Laurentz.

C. lurida, Wahl. Pale Sedge.   (*C. tentacula*, Muhl.)
Swampy spots. Monongalia: along Falling Run. Wood: near Lockhart's Run. Fayette: near Nuttallburg—L. W. N.

Var. **gracilis.** Bailey. Slender, Pale Sedge.
On mossy boulders. Webster: along Buffalo Bull Mountain, alt. 2575 feet. Tucker: near Falls of Blackwater.

C. stenolepis, Torr.
Damp meadows. Monongalia: along Falling Run. Fayette: near Nuttallburg—L. W. N.

C. squarrosa, Dewey.
River banks. Monongalia: mouth of Falling Run. Fayette: near Nuttallburg—L. W. N.

C. stricta. Lam.
Sphagnum Swamp. Fayette: near Nuttallburg—L. W. N.

C. torta, Boott.
Cold damp places. Fayette: near Nuttallburg—L.W.N.

C. prasina, Wahl.
Fayette: bed of Keeney's Creek, near Nuttallburg—L. W. N.

C. crinita. Lam.   L. W. N., M. & G.
Damp swales. General throughout the State even in the higher mountains.

C. CRINITA X C. PRASINA ? Bailey.
Fayette: along a run in Sugar Camp Hollow—L.W.N.

C. virescens. Muhl.
Banks and copses. Wood: near Lockhart's Run.

Var. **costata,** Dewey.
Banks. Fayette: near Nuttallburg—L. W. N.

C. triceps, Michx., *var.* **hirsuta,** Bailey.
>Dry pastures. Wood: near Kanawha Station. Fay
ette: woodland border of swamp near Nuttallburg—L. W. N.

C. debilis, Michx.. *var.* **Rudgei,** Bailey.
>On mossy boulders. Randolph: summit of Rich
Mountain, alt. 2850 feet; undrained meadows of Tygart's
Valley River: Point Mountain, on perfectly dry rock, alt.
3650 ft.
>A very common sedge in the mountains on mossy
boulders and along runs. Beautiful growths occur all along
the Blackwater Fork of Cheat. Fayette: near Nuttallburg—
L. W. N.

C. venusta, Dewey. *var.* **minor,** Boeckl.
>Fayette: near Masterton's swamp. Nuttallburg—L.
W. N.

C. gracillima, Schw.
>Low grounds. Tucker: near Falls of Blackwater.
Fayette: near Nuttallburg—L. W. N.

C. grisea, Wahl.. *var.* **angustifolia.** Boott.
>Fayette: on banks near Nuttallburg—L. W. N:

C. glaucodea. Tuck.
>Meadows: Fayette near Nuttallburg—L. W. N.

C. laxiflora, Lam.
>Ohio: on Thomas Hill near Wheeling.

Var. **latifolia.** Boott.
>Deep woods. Monongalia: banks of Day Creek.
Wirt: banks of Straight Creek. Fayette: near Nuttallburg
—L. W. N.

Var. **patulifolia,** Carey.
>Fayette: shady bank, near Nuttallburg—L. W.. N.

C. digitalis. Willd.
>In deep woods. Grant: near Bayard. Fayette: near
Nuttallburg—L. W. N.

C. laxiculmis, Schw.
>Glady places. Fayette: near Nuttallburg—L. W. N.

C. plantaginea, Lam.
>Rich wood. Fayette: near Nuttallburg—L. W. N.

C. varia *var.* **colorata.** Bailey.
 Fayette: near Nuttallburg—L. W. N.

C. Pennsylvanica, Lam.
 Ohio: near Wheeling —M. & G.

C. communis. Bailey.
 Fayette: dry soil under cliffs, near Nuttallburg—L.
W. N.

C. Jamesii, Schwein.
 Fayette: open woods near Nuttallburg—L. W. N.

C. polytrichoides. Muhl.
 Fayette: sphagnum swamp near Nuttallburg—L.
W. N.

C. Fraseri. And. Frazer's Sedge.
 The following remarks of Prof. T. C. Porter render it
evident that this rare and odd sedge came originally from
Randolph or Barbour county, each of which lies between the
headwaters of the two Kanawhas:
 "Muhlenberg, in his *Descriptio uberior Graminum*, etc., p.
265, under C. lagopus ?, which is C. Fraseri, Andrews, adds
these words, 'Habitat in Tyger-Valley, Pennsylvaniae, *unde
siccum habeo et circum.*' Kin, the German gardner who collected in Pennsylvania, brought it home, and his label reads
thus: 'Deigher Walli in der Wilternus.' Dr. Gray has
shrewdly conjectured that by 'Deigher Walli,' or Tygert Valley, is meant Tygart's Valley, which lies further south in
Virginia." In a foot-note, Prof. Porter adds "a box containing the Carices of Muhlenberg has just been discovered (1877)
in the herbarium of the Academy, Philadelphia, and the label
attached to the specimens of Kin's collection places Tyger
Valley 'prope amnem Kenahway.' "
 As the two Kanawhas lie entirely within the State of
West Virginia, there seems to be little doubt as to the pro-
priety of including this species in this Flora.
 Since writing the above Mr. Nuttall has found a plen-
tiful station for this species near Nuttallburg in the Great
Kanawha region.

C. stipata. Muhl.
 Ohio: near Wheeling M. & G. Fayette: near Nuttallburg L. W. N. Mercer: near Bluefield.

C. vulpinoidea, Michx.
 Monongalia: along Falling Run. Wirt: near Burning Spring. Fayette: near Nuttallburg—L. W. N.

C. rosea. Schk.
> Fayette R. R. bank near Nuttallburg—L. W. N.

Var. **radiata,** Dewey.
> Open places. Monongalia: near Round Bottoms.
> Wirt: along Straight Creek. Randolph: on Point Mountain,
> alt. 3450 feet; also on a dry rock, alt. 3650 ft. Fayette:
> cliffs, rocks and banks, near Nuttallburg—L. W. N.

Var. **retroflexa,** Torr.
> Fayette: Swampy place near Nuttallburg—L. W. N.

C. sparganioides. Muhl.
> Fayette: Wet open banks near Nuttallburg—L. W. N.

C. Muhlenbergii. Schk., *var.* ———
> An intermediate between the type and variety *enervis*,
> *fide* Bailey. Randolph: on an undrained meadow along
> Tygart's Valley River, near Cricard.

Var. **enervis.** Boott.
> Opens. Lewis: along Leading Creek.

C. cephalophora. Muhl.
> Fayette: banks near Nuttallburg—L. W. N.

C. canescens, L., *var.* **vulgaris,** Bailey.
> On a dry conglomerate rock. Randolph: summit of
> Rich Mountain, alt. 2850 feet.

C. tribuloides, Wahl.
> In a springy rill. Wood: near Kanawha Station.

Var. **turbata,** Bailey.
> Fayette: low copse near Nuttallburg—L. W. N.

C. scoparia, Schk.
> Open swales. Monongalia: along Falling Run.
> Fayette: banks of river and in sphagnum bog near Nuttall-
> burg—L. W. N.; dry glade, Alderson Farm—L. W. N.

C. straminea. Wild.
> Dry soils. Fayette: near Nuttallburg—L. W. N.

GRAMINEÆ.

> All doubtful forms have been kindly identified by Dr.
> Geo. Vasey.

SPARTINA, Schreb.

S. cynosuroides. Willd. Fresh-water Cord-grass.
Ohio: on Bogg's Island—M. & G. Fayette: near
Nuttallburg—L. W. N.

PASPALUM. L.

P. setaceum, Michx.
Sandy soils. Monongalia: at the end of the Hog
Back, Decker's Creek near Morgantown.

Var. **ciliatifolium**(Michx.)
Fayette: near Nuttallburg—L. W. N.

P. læve, Michx.
Fayette: near Nuttallburg—L. W. N.

PANICUM. L.

P. SANGUINALE. L. Crab, or Crow-foot Gress. L. W. N.
Cultivated and waste grounds. Common throughout
the State.

forma **depauperata,** Vasey.
Dry sterile fields. Monongalia: up Falling Run beyond the campus.

P. proliferum, Lam. M. & G.
River banks. Monongalia: banks of the Monongahela, near Uffington and Morgantown. Fayette: near Nuttallburg— L. W. N.

Var. **geniculatum,** Ell., Vasey. (P. *geniculatum*, Ell.)
Waste grounds. Monongalia: near Morgantown.
Fayette: near Nuttallburg—L. W. N.

P. capillare. L. "Tickle Grass." M. & G.
Dry Fields. Monongalia: near Morgantown. Greenbrier: near White Sulphur Springs. Fayette: near Nuttallburg—L. W. N.

Var. **campestre.** Gattinger.
Dry fields. Monongalia: on a sterile hillside up Falling Run beyond the Campus.

P. anceps. Michx.
Fayette: near Nuttallburg—L. W. N.

465

P. agrostoides, Muhl.

Wet meadows. Kanawha: near Allen's Fork. Monongalia: below the mouth of Falling Run. Upshur: in a damp meadow near Lorentz. Fayette: near Nuttallburg.— L. W. N.

P. virgatum, L.

Moist sandy soil. Monongalia: streets of Morgantown, growing between the bricks of walks, and seeming to flourish in direct proportion to the amount it is trodden upon. Mason: banks of the Ohio River near Point Pleasant. Fayette: near Nuttallburg—L. W. N.

P. latifolium, L.

Moist thickets. Wirt: near Burning Springs. Greenbrier: near White Sulphur Springs. Monongalia: near Little Falls. Fayette: near Nuttallburg—L. W. N.

P. clandestinum, L. "Deer-tongue Grass."

Damp meadows. Monongalia: common throughout. Upshur: near School House Summit. Summers: near Rifle. Monroe: near Wolf Creek. Fayette: near Nuttallburg.

A very nutritious grass; especially enjoyed by horses. Our analysis gives it a nutrative ratio of 1:1.24.

P. depauperatum, Muhl.

Dry opens. Fayette: near Nuttallburg—L. W. N. Monongalia: at Roundbottoms.

P. dichotomum, L.   L. W. N.

Common everywhere, throughout the State.

forma **commune**, Man.

Common, especially along river banks: Monongalia and Marion: along the Monongahela River. Fayette: near Nuttallburg.

forma **fasciculatum**, Man.

Drier situations. Monongalia and Marion: along the F. M. & P. R. R.

forma **gracile**, Man.

Wet bottoms, usually along runs. Wood: in a swampy ditch near Kanawha Station. Wirt: in a weedy ditch near Reedy Ripple; in a spring rill in rich woods near Straight Creek. Randolph: in a cold rill in deep woods, on Point Mountain, alt. 3,200 ft.

Var. **elatum.**
>Monongalia: along the F. M. & P. R. R., below Little Falls, in a sandy ditch. Fayette: near Nuttallburg—L. W. N.

P. pubescens. Lam.
>Damp places. Lewis: along Leading Creek.

P. CRUS-GALLI. L.   Barnyard Grass.   L. W. N.
>Ditches and waste grounds. Monongalia: near Morgantown. Greenbrier: near White Sulphur Springs. Summers: near Hinton. Fayette: near Nuttallburg.

Var. **hispidum** Muhl., Torr.
>Ditches. Monongalia: along the F. M. & P. R. R., near Little Falls.

CHAMÆRAPHIS. R. Br. (1810).
(Setaria, Beauv. 1812.)

C. GLAUCA (L.).   Fox-tail Grass.   L. W. N., M. & G.
>Common throughout the State, especially in stubble fields.

C. VIRIDIS (L.).   L. W. N., M. & G.
>Roadsides and cultivated fields. Jackson: near Sandyville, and on Limestone Ridge.

C. ITALICA (L.).
>Fayette: near Nuttallburg- L. W. N.

CENCHRUS. L.

C. tribuloides. L.   Hedgehog Grass.   Bur-grass.
>Jefferson: near Harper's Ferry -M. & G.

HOMALOCENCHRUS. Mieg (1768.)
(Leersia, Swartz. 1788.)

H. Virginica (Willd.), Britt.   White Grass.   (*Leersia Virginica,* Willd.)
>Wet places. Monongalia: near Beechwoods and Little Falls. Fayette: near Nuttallburg- L. W. N.

H. oryzoides (L.), Poll.   Rice Cut-grass.   (*Leersia oryzoides,* Swartz.)
>Wet grounds Nicholas: in Collett's Glade. Fayette: near Nuttallburg -L. W. N.

TRIPSACUM, L.

T. dactyloides. L.

Moist places. Fayette: near Nuttallburg—L. W. N.

ANDROPOGON, Royen.

A. provincialis, Lam. Beard Grass. (*A. furcatus.* Muhl.)
Damp places. Randolph: along Tygart's Valley River
near Beverly. Monongalia: along the Monongahela below
Morgantown. Fayette: near Kanawha Falls —James: near
Nuttallburg—L. W. N. Summers: near Greenbrier Stock-
yards; and along Greenbrier River. Taylor: near Grafton.

A. scoparius. Michx. "Broom Sedge."
Dry soils. Monongalia: about Morgantown, plentiful.
Webster: near Upper Glade. Mason: near Point Pleasant.
Taylor: near Grafton, and in every county visited.

This species, which threatens to be our most perni-
cious and wide-spread weed, is advancing eastward with the
utmost aggressiveness. It has absolutely no fodder qualities,
its nutritive ratio being only 1 : 14.50, and its value as a fer-
tilizer only $2.37 per dry ton. The method of combatting
this pest is as yet a mere matter of conjecture.

The plant is reported from Brooke: near Wellsburg
and Fowler's. Barbour: near Overfield, Pepper and Philippi.
Braxton: near Bulltown, Frametown, Tate Creek, Elmyra
and Newville. Cabell: near Union Ridge and Barboursville.
Clay: near Valley Fork. Doddridge: near Leopold, New
Milton and Center Point. Fayette: near Fayetteville, Moun-
tain Cove and Beets. Grant: near Medley, Petersburg and
Greenland. Greenbrier: near Frankford, Trout Valley and
Fort Spring. Hampshire: near Shanesville, Springfield and
Three Churches. Harrison: near Clarksburg, Lost Creek,
Bridgeport, Wallace, Wilsonburg, Adamsville, Good
Hope and Mount Clare. Hardy: near Moorefield
and Wardensville. Jefferson: near Kabletown. Jack-
son: near Grass Lick, Sandy, Wilding, Belgrove
Odaville, Silverton and Kentuck. Kanawha: near Pocatali-
go, Blandon and Gazil. Lewis: near Vadis, Camden, Walk-
ersville and Aberdeen. Lincoln: near Hamlin. Mercer:
near Princeton and New Hope. McDowell: near Squire
Jim. Monroe: near Cashmere and Johnson's X Roads.
Mason: near New Haven, Grimm 's Landing and Maggie.
Marion: near Canton, Farmington, Eldora, Barrackville,
Boult and Gray's Flat. Monongalia: near Morgantown,
Putnam: near Hurricane, Paradise, Carpenters and Confi-
dence. Pleasants: near New Hebron. Preston: near Ma-
sontown, Eglon and Amblersburg. Ritchie: near Ritchie
C. H. Berea and Cornwallis. Randolph: near Florence and

Lee Bell. Roane: near Pencil, Newton, Looneyville, Clio, Reedy and Countsville. Raleigh: near Table Rock, Egeria and Raleigh C. H. Summers: near Forest Hill, Talcott, Clayton and Indian Mills. Taylor: near Knottsville, Thornton, Grafton and Meadland. Tyler: near Wick and Long Reach. Tucker: near Texas. St. Georges and Hendricks. Upshur: near Lawrence, Overhill, Evergreen, French Creek and Hemlock. Wayne: near Adkin's Mills and Egypt. Wetzel: near New Martinsville. Wirt: near Elizabeth, Burning Springs, Reedy Ripple and Lee. Wood: near Blennerhassett and Kanawha Station. Webster: near Replete, and Welch Glade.

A. Virginicus, L.
Fayette: near Nuttallburg—L. W. N.

CHRYSOPOGON, Trin.

C. nutans(L.). Benth. Indian Grass. Wood Grass.
Fayette: near Nuttallburg—L. W. N.

PHLARIS, L.

P. arundinacea, L. Reed Canary-grass.
Wet places. Wood. in a spring rill near Kanawha Station.

ARISTIDA, L.

A. dichotoma, L. Poverty Grass.
Fayette: near Nuttallburg—L. W. N.

A. oligantha, Michx. Triple-awned Grass.
Dry banks. Kanawha: near Charleston—Barnes.

A. lanata, Poir.
Fayette: near Nuttallburg—L. W. N.

MUHLENBERGIA, Schreber.

M. sobolifera(Muhl.). Trin.
Fayette: near Nuttallburg—L. W. N.

M. Mexicana(L.). Trin.
Fayette: near Nuttallburg—L. W. N. Monongalia: on campus, Morgantown.

M sylvatica Torr. . Torr. & Gray.
Fayette: near Nuttallburg—L. W. N.

M. diffusa, Schreb. Nimble Will.
Dry ground. Monongalia: along Decker's Creek.
Fayette: Nuttallburg—L. W. N.

BRACHYELYTRUM, Beauv.

B. aristosum(Michx.), B. S. P. (*B. aristatum*. Beauv.)
Rocky woods. Webster: near Welch Glade. Tucker:
near the Falls of Blackwater. Fayette: near Nuttallburg—
L. W. N.

Var. **glabratum**. Vasey, MSS.
A new variety discovered by Mr. L. W. Nuttall. It
agrees with the species except that it is perfectly smooth, and
has an awn pointed second glume which is about one-half the
length of the flowering glume. Its most striking peculiarity
is that it has invariably two culms from each rootstock.
High, rocky woods. Fayette: near Nuttallburg, rare
—L. W. N.

PHLEUM, L.

P. PRATENSE. L. Timothy. L. W. N., M. & G.
A common escape from cultivation throughout the
State. Found even in the forests of the highest Alleghanies.

AGROSTIS, L.

A. ALBA, L. White Bent-grass.
Meadows and fields. A frequent escape in the western
counties. Fayette: near Nuttallburg - L. W. N.

Var. **VULGARIS**(With.). Thurb. Red Top. L. W. N.
Meadows the fields. Common throughout the State.

Forma, **artistata**.
Damp meadowlands. Monongalia: along Falling Run

A. prennans(Walt.). Tuckerm. Thin Grass.
Damp, shady places. Randolph: along Cheat River
in clearings. Monongalia: near Little Falls. Fayette: near
Nuttallburg—L. W. N.

A. hiemalis(Walt.), B. S. P. Hair Grass. *A. scabra*, Willd.
Moist fields. Preston: near Terra Alta. Fayette:
near Nuttallburg—L. W. N

CINNA. L.

C. arundinacea, L. Wood Reed-grass.
Wet places. Randolph: along Tygart's Valley River
near Huttonsville. Fayette: near Nuttallburg—L. W. N.

ARRHENATHERUM, Beauv.

A. ELATIUS L., Mert. & Koch. Oat Grass. (A. avenaceum, Beauv.)
Fields and yards. Monongalia, abundant and becoming a nuisance.
The lower campus (half orchard) that yielded a good
crop of Orchard-grass last season, was almost completely
this species this year. Our analysis of this grass shows a
nutritive ratio of only 1:8.13. Fayette: near Nuttallburg—
L. W. N.

HOLCUS, L.

H. LANATUS. L. Velvet Grass. "Old White Top." "Feather
Grass."
Frequent in damp meadows. Monongalia: along the
F. M. & P. R. R. Upshur: near Laurentz. Randolph:
along Tygart's Valley River. Grant: near Bayard. Nicholas:
in Collett's Glade. Fayette: near Hawk's Nest, and Kanawha Falls—James: Nuttallburg—L. W. N.

DANTHONIA. DC.

D. spicata(L.). Beauv. Wild Oat-grass.
Dry soil. Monongalia: near Beechwoods and Little
Falls. Randolph: on a dry boulder, summit of Point Mountain, alt. 3750 ft. Fayette: near Nuttallburg—L. W. N.

D. compressa. Austin.
Dry banks. Tucker: near the Falls of Blackwater.
Fayette: near Nuttallburg—L. W. N.

ELEUSINE, Gærtn.

E. INDICA(L.). Gaertn. Dog's Tail or Wire Grass.
Yards and streets. Kanawha: near Cannellton. Monongalia: near Morgantown. Mason: near Point Pleasant.
Berkeley: near Martinsburg. Fayette: near Nuttallburg—
L. W. N.

E. ÆGYPTICA, Pers. Crab-grass. Yard-grass.
Yards and lawns. Monongalia: on the campus.

SIEGLINGIA, Bernh. 1800
(Triodia, R. Br. 1810)

S. cuprea(Michx.) Tall Red Top.
Dry or sandy fields. Monongalia: near Little Falls.
Wood: near Selden. Fayette: near Nuttallburg—L. W. X.

EATONIA. Raf.

E. obtusata(Michx.), Gray.
Rich woods. Fayette: near Nuttallburg—L. W. X.

E. Pennsylvanica(Spreng), Gray.
Moist opens. Monongalia: near Beechwoods and in
the campus.

ERAGROSTIS. Beauv.

E. hypnoides (Lam.), B. S. P.   *E. reptans*, Nees.
Shores of rivers. Summers: along New River near
Hinton. Wood: along the Ohio River near Parkersburg.
Mason: near Point Pleasant. Ohio: near Wheeling—
M. & G.

E. MINOR, Host.   *E. poaoides*, Beauv.
Waste places. Mason: near Point Pleasant.

E. MAJOR. Host.
Fayette: near Nuttallburg—L. W. X.

E. PILOSA(L.), Beauv.
Jefferson: near Harper's Ferry—M. & G. Fayette:
near Nuttallburg—L. W. X.

E. Purshii. Schrader.
Sterile or sandy soils. Monongalia: near Little
Falls, and on the campus.

E, capillaris,(L.), Nees.
Fayette: near Nuttallburg—L. W. X.

E. Frankii, Meyer.
Shores of rivers. Summers: along New River near
Hinton.

MELICA. L.

M. mutica, Walt.
Rich soil. Fayette: near Nuttallburg—L. W. X.

CORYCARPUS, Zea. (1806)
(Diarrhena. Beauv. 1812)

C. Americana(Beauv.) Kuentz.
Shaded river banks. Fayette: near Nuttallburg—L.
W. N.

UNIOLA, L.

U. latifolia, Michx.
Shady places: Fayette: near Nuttallburg—L. W. N.

DACTYLIS, L.

D. GLOMERATA. L. Orchard Grass.   L. W. N.
Fields and meadows. Common throughout, especially in shady places.

POA, L.

P. ANNUA, L. Low Spear-grass.
Ohio: near Wheeling—M. & G. Fayette: near Nuttallburg—L. W. N.

P. COMPRESSA, L. Wire Grass.   L. W. N.
Sterile soil in crevices of rocks. Plentiful throughout
the State.

Forma depauperata.
On rocks. Monongalia: along Falling Run, especially
at the cascade.

P. pratensis, L. Blue Grass.   L. W. N.
Dry soils and meadows. Common throughout the
State, even in the higher Alleghanies.

P. TRIVALIS. L. Roughish Meadow-grass.
Moist meadows. Monongalia: along the Monongahela at Little Falls.

P. sylvestris, Gray.
Ohio: near Wheeling - M. & G.

P. alsodes, Gray.
Hill-side woods. Monongalia: along Day Creek near
Little Falls. Fayette: near Nuttallburg—L. W. N.

P. flexuosa. Muhl. (not Wahl.).
Tyler: near Long Reach.

P. brevifolia, Muhl.
>Ohio: Woods Run near Wheeling—M. & G. Fayette:
near Nuttallburg—L. W. N.

PANICULARIA, Fabr. (1763)
(Glyceria, R. Br. 1810.)

P. Canadensis Michx.).
>Woods. Fayette: near Nuttallburg—L. W. N.

P. elongata(Torr.). Manna Grass.
>Wet woods. Webster: along the ridge of Buffalo Bull
Mountain. Fayette: near Nuttallburg—L. W. N.

P. nervata(Willd.). Foul Meadow-grass.   L. W. N.
>Moist meadows. Common throughout.

forma. **major.**
>Monongalia: sand bars in Monongahela River near
Little Falls.

FESTUCA, L.

F. octoflora, Walt.   (F. tenella, Willd.)
>Dry open woods. Wirt: hills above Burning Springs.

F. ELATIOR. L. Meadow Fescue.
>Fields. Monongalia: near Beechwoods. Fayette:
near Kanawha River—James: near Nuttallburg—L. W. N.

F. pratensis, Huds.
>Meadows. Wood: near Kanawha Station; and else-
where frequent.

F. nutans, Spreng.
>Rocky woods. Randolph: on dry boulder, summit
of Rich Mountain. Webster: roadsides near Long Glade.
Fayette: near Nuttallburg—L. W. N.

BROMUS, L.

B. MOLLIS. L. Soft Chess.
>Wheat fields and waste grounds. Monongalia: on
the campus.

B. SECALINUS. L. Cheat or Chess.   L.W.N. M.&G.
>Wheat fields and waste places. A too common nuisance.

B. RACEMOSUS. L. Upright Chess.   L. W. N.
>Fields common throughout the State.

B. ciliatus, L..

River banks. Wood: along the Little Kanawha River near Kanawha Station. Monongalia: along the Monongahela near Little Falls.

LOLIUM, L.

L. PERENNE. L. Darnel. Rye Grass. English Blue Grass. Lawns, introduced with "Lawn Grass Seed". Monongalia: at Morgantown. Fayette: near Nuttallburg—L.W.N.

ELYMUS, L.

E. Virginicus, L.. Virginia Wild-rye.
River banks. Ohio: near Wheeling—M. & G. Mason: near Point Pleasant. Fayette: near Nuttallburg—L. W. N.

E. Canadensis. L... *var.* **glaucifolius** (Muhl.), Gray.
Dry banks and roadsides. Kanawha: along 8-Mile Creek Mason: near Point Pleasant. Fayette: near Nuttallburg—L. W. N.

E. striatus, Willd.
Rocky banks. Monongalia: along the Monongahela river below Morgantown.

Var. **villosus.** Gray.
Monongalia: banks of the Monongahela below Morgantown.

HYSTRIX, Moench. (1794.)
(Asprella, Willd. 1809.)

H. Hystrix (L.) Bottle-rush Grass. L. W. N.
Moist woodland banks. Scattering in Wood, Wirt, Calhoun, Gilmer, Lewis, Upshur, Monongalia, Randolph, Webster and Nicholas counties. Fayette: along the Gauley River: near Gauley Mountains: near Nuttallburg: near Kanawha Falls—James. Summers: near Hinton. Mason: near Point Pleasant. Harrison: near Lumberport.

GYMNOSPERMÆ.

CONIFERÆ.

· THUJA, L.

T. occidentalis, L.. Arbor Vitæ.
Dry, rocky hills. Mineral: on Knobby Mountains.
Grant: near Petersburg.

JUNIPERUS. L.

J. communis, L.. Juniper.
Dry sterile hills. Wood. near Kanawha Station.
Mineral: on Knobby Mountains. Fayette, near Nuttallburg
—L. W. N.

J. Virginiana, L.. Red Cedar. Savin. M. & G.
Wood: near Lockhart's Run and Kanawha Station.
Wirt: on Nigh-cut Hill. Fayette: near Crescent Kanawha
Falls—James; Nuttallburg—L. W. N. Mineral: on Knobby
Mountains. Jefferson: along the Potomac. Mason: near
Point Pleasant. Berkeley: near Martinsburg. Monongalia:
near Morgantown. Mercer: near Ingleside. Jackson and
Mason: along the Ohio River: Cabell: along the C. & O. R.
R., from Huntington, to St. Albans in Kanawha county.

TAXUS, L.

T. baccata, L... *var.* **Canadensis** Willd., Gray. "Creeping Hemlock." Yew.
Damp hillside woods. Marion: near the mouth of
· Buffalo Creek—K. D. Walker. Taylor: near Nuzums. Fayette: along Williams River.

PINUS. L.

P. strobus, L.. White Pine.
Wood: near Leachtown. Wirt. near Burning Spring.
Calhoun: near White Pine and Laurel Run. Pocahontas:
near Sunset. Greenbrier: near Caldwell. Fayette: near
Nuttallburg, probably introduced L. W. N.

P. Taeda, L.. Loblolly, or Old-field Pine.
Opens. Wood: near Kanawha Station. Mineral.
Hampshire, and Hardy along the table-lands.

P. rigida. Mill. Pitch Pine.
Kanawha: near Charleston—Barnes. Fayette: near Nuttallburg, alt. 2000 ft.—L. W. N. Greenbrier: near White Sulphur Springs. Mineral: on Knobby Mountains.

P. pungens. Michx. f. Table Mountain Pine. M. & G.
Dry rocky soils. Kanawha: near Charleston—Barnes. Mineral on Knobby Mountain. Pendleton: foot hills of Spruce Knob—A. D. Hopkins.

P. Virginiana, Mill. Scrub Pine. (*P. inops.* Ait)
Sterile hills. Greenbrier: near Caldwell. Fayette: near Nuttallburg, a common second growth—L. W. N. Berkeley: near Martinsburg. Mercer: near Ingleside.

P. echinata. Mill. Yellow Pine. (*P. mitis.* Michx.)
Wood: near Leachtown. Randolph: near Valley Bend. Summers: near Hinton.

PICEA, Link.

P. Mariana(Mill.) B. S. P. Black Spruce. "Yew Pine." "White Spruce." "Spruce Pine." *Abies nigra,* Poir.
Magnificent forests in the following regions where it grows at elevations varying from from 2500 to 4000 ft. :

	Acres.
Randolph County on Elk and Gauley waters	15,000
Randolph County on Cheat waters	120,000
Randolph County on Mill Creek	5,000
Randolph County on Elk Mountain	500
Pocahontas County on Shaver's Fork of Cheat	20,000
Pocahontas County on the headwaters of Greenbrier River.	100,000
Pocahontas County on Elk and Gauley headwaters	100,000
Tucker County on Cheat waters.	50,000
Mineral County	25,000
Greenbrier County (by actual survey)	35,499
Total acreage	470,999

This estimate will probably fall under the actual amount.
Summers: along Greenbrier river near Talcott. Mercer: near Ingleside.

TSUGA. Carrier.

T. Canadensis(L.), Carr. Hemlock. Hemlock Spruce. *Abies Canadensis.* Michx. M. & G.
Rocky woods. Wirt: along Straight Creek. Calhoun: on Laurel Run. Nicholas: near Beaver Mills. Grant: near Bayard, abundant. Monroe: near Alderson. Preston: near Terra Alta. Fayette: near Nuttallburg, plentiful—L. W. N. Mercer: near Ingleside and Princeton. McDowell: near Elkhorn. Monongalia: near Uffington and Tibb's Run.

ABIES, Miller.

A. balsamea, Miller. "Blister Pine." Balm-of-Gilead Fir, Balsam Fir. Mountain Swamps. Randolph: about two miles beyond Cheat Bridge along the Staunton Pike.

SELAGINELLEÆ.

SELAGINELLA, Beauv.

S. rupestris(L.), Spring.*
Jefferson: near Harper's Ferry M. & G.

LYCOPODIACEÆ.

LYCOPODIUM. L.

L. lucidulum, Michx.   M. & G.
Cold, damp woods Grant: near Bayard. Randolph: near Cheat Bridge. Gilmer: near Glenville—V. M. Fayette: near Nuttallburg. alt. 2000 ft.—L. W. N. Tucker: near Davis.

L. annotium. L.
Cold woods. Randolph: plentiful along the west slope of Cheat Mountains near Cheat Bridge. Fayette: near Nuttallburg—L. W. N.

L. obscurum, L. Ground Pine. (*L. dendroideum*, Michx.)M.&G. Deep, moist woods. With the last, plentiful.

L. clavatum, L. Club Moss.
Dry woods. Common throughout the State.

L. complanatum, L. Trailing Christmas Green.   M. & G.
Deep coniferous woods. In the Alleghanies of Grant, Tucker, Randolph and Pocahontas counties. Fayette: near Nuttallburg—L. W. N.

OPHIOGLOSSACEÆ

OPHIOGLOSSUM, L.

O. vulgatum. L. Adder's Tongue.
Wet meadows and woods. Gilmer: near Glenville— V. M.

BOTRYCHIUM. Sw.

B. ternatum Thunb.). Sw. Moon-wort.
Dry woods. Monongalia. Marion. Preston: in Laurel
hills. Grant: near Bayard. Gilmer: near Glenville—V. M.
Fayette: near Nuttallburg. glades, alt. 2000 ft.- L. W. N.

Var. **australe**(R. Br.). Eaton.
Grassy places. Monongalia: on the campus near
Morgantown.

Var. **rutæfolium**. Man.
Rich opens. McDowell: near Elkhorn.

Var. **obliquum**(Muhl.), Milde.
Fayette: near Nuttallburg. dry opens, alt. 2000 ft.—
L. W. N. Ohio: hills back of Wheeling—M. & G. Monon-
galia: on the campus near Morgantown.

Var **dissectum**(Spreng.), Milde.
Glades. Fayette: near Nuttallburg, alt. 2000 ft.—L.
W. N. Monongalia: on the campus near Morgantown.

B. Virginianum (L.). Sw. "Indicator." M. & G.
Rich woods. Grant: near Bayard. Randolph: on
Rich Mountains. Monongalia: Cheat river near Camp Eden.
Gilmer: near Glenville. Jackson: near Ripley. where it is
often called "Indicator" as its growth is thought to indicate
the presence of Ginseng in the locality.

—————()————

PTERIDOPHYTA.

FILICES.

POLYPODIUM, L.

P. vulgare, L. Common Polypody. M. & G.
Common on mossy rocks and in rocky woods. Kanawha: near Charleston—Barnes: near Coalburg—James.
Gilmer: near Glenville V. M. Fayette: near Nuttallburg
L. W. N. Grant: near Bayard and along Buffalo Creek,
Monongalia: along Cheat River. Tucker: along Beaver
Creek and Blackwater. Randolph: on Rich and Cheat Mountains. Greenbrier: near White Sulphur Springs. Summers: near Hinton. McDowell: near Elkhorn.

Forma **biserrata,** mihi.
A form found upon mossy boulders along the Blackwater Fork of Cheat: with regularly doubly-serrate divisions of the thinish fronds.

P. incanum. Sw.
Rocks and tree trunks. Fayette: near Hawk's Nest
—Porter.

CHEILANTHES. Sw.

C. vestita (Spreng.). Sw. Lip Fern.
Rocky places. Jefferson: near Harper's Ferry—M. & G.

C. lanuginosa, Nutt. Wooly Lip-fern.
Cliffs. Fayette: near Kanawha Falls—James.

PELLÆA. Link

P. atropurpurea (L.), Link. Cliff Brake. M. & G.
Dry rocks. Fayette: near Nuttallburg, rare L. W.
N. Mercer: near Beaver Spring on exposed ledges, plentiful.

PTERIS, L.

P. aquilina. L. Brake or Bracken. M. & G.
Thickets and hillsides. Common throughout. Upshur: at School-house Summit. Webster: in Upper Glade.

Kanawha: near Coalburg and Charleston- James. Fayette:
near Nuttallburg—L. W. N.

ADIANTUM, L.

E. pedatum, L. Maiden Hair Fern.　　L. W. N.. V. M.. M. & G.
Rich moist woods. Common throughout the State.

ASPLENIUM, L.

A. pinnatifidum(Muhl), Nutt. Spleenwort.
　　　Cliffs and rocks. Jefferson: near Harper's Ferry —M.
& G. Fayette: near Nuttallburg, in clefts of boulders, rare
—L. W. N.

A. Trichomanes, L.　　　　　　　　　　　M. & G.
　　　Shaded cliffs. Wirt: near Burning Springs. Nicho-
las: along Peter Creek. Fayette: near Nuttallburg—L. W.
N.: and along the Gauley River. Kanawha: near Coalburg
—James. Gilmer: near Glenville. Greenbrier: near White
Sulphur Springs.

A. platyneuron (L.). Oakes.　　　　　　　(A. ebenum. Ait.)
　　　Frequent in rocky open woods. Kanawha: near
Charleston--Barnes. Fayette: near Nuttallburg—L. W. N.
Wirt: near Burning Springs.

A. montanum. Willd.
　　　Cliffs and rocks. Fayette: near Hawk's Nest- Porter (see Meehan's Monthly, Aug. 1892, plate): near Nuttallburg, alt. 2000 ft. L. W. N. Jefferson: near Harper's Ferry M. & G. Randolph: near Helvetia M. & G. Grant:
near Bayard. Monongalia: on boulders along Tibb's Run,
plentiful.

A. angustifolium, Michx.
　　　Rich woods Ohio: near Wheeling M. & G. Fayette: near Nuttallburg- L. W. N.

A. acrostichoides. Sw.　　　　　　　(A. thelypteroides. Michx.)
　　　Rich woods. Kanawha: near Charleston—James.
Fayette: near Nuttallburg—L. W. N. Jefferson: near Har-
per's Ferry--M. & G.

A. Filix-fœmina L., Bernh.
　　　Moist woods. Gilmer: near Glenville- V. M Fay-
ette: near Nuttallburg- L. W. N. Ohio: near Wheeling
M. & G. Tucker: near the Falls of Blackwater.

CAMPTOSORUS, Link.

C. rhizophyllus L.., Link. Walking Fern. Walking Leaf.
Shaded rocks and conglomerate boulders. Wirt: near
Burning Springs. Fayette: along the Gauley River; near
Kanawha Falls —James. Tucker: at Blackwater Falls.
Gilmer: near Glenville –V. M. Kanawha: near Coalburg –
James. Fayette: near Nuttallburg –L. W. X. Monongalia:
near Morgantown.

PHEGOPTERIS. Fee.

P. connectilis Michx.), B. S. P. Beech Fern. *P. polypodioides.* Fee.
Damp woods. Gilmer: near Glenville—V. M. Tuck-
er: near the Falls of Blackwater

P. hexagonoptera L.). Fee.
Open woods. Gilmer: near Glenville V. M. Fay-
ette: near Nuttallburg, shaded fence rows and deep woods—
L. W. X. Ohio: near Wheeling—M. & G.

P. Dryopteris(L.), Fee.
Rocky woods. Preston: near Rowlesburg—M. & G.

ASPIDIUM, Sw.

A. Thelypteris L.). Sw. Shield-fern.
Marshy Meadows. Ohio: near Wheeling—M. & G.

A. Noveboracense L.). Sw. New York Shield Fern. M. & G.
Moist woods. Randolph: on Rich Mountain, alt.
1850 feet. Fayette: near Nuttallburg—L. W. X.; near Kanawha Falls James. Kanawha: near Charleston--James.

A. fragrans. Swartz.
Opens. Pocahontas: near summit of Spruce Knob,
alt: 4800 ft.. where it is cut and cured for fodder.

A. spinulosum, Sw. Wood-fern.
Damp woods. Wirt: above Burning Springs. Ran-
dolph: on Rich Mountain. Fayette: near Nuttallburg—L.
W. X. Preston: near Terra Alta. McDowell: near Elkhorn.

Var. ——
In deep, wet woods under Black Spruce. Randolph:
near Cheat Bridge, and Shades of Death.

Var. **intermedium** Willd. Eaton. Common Wood-fern.
Deep rich woods throughout Grant, Tucker, Randolph
and Pocahontas Counties. Upshur: near Beech and Middle
Fork. Fayette: near Nuttallburg--L. W. X.

Var. **dilatatum** (Sw.), Hook.
> Deep woods. Ohio near Wheeling—M. & G.

A. cristatum (L.) Sw.
> Swampy places. Preston; near Cranberry Summit—M. & G.; near Reedsville and Terra Alta.

A. Goldianum, Hook.
> Rich moist woods. Preston; near Cranberry Summit M. & G. Fayette; near Nuttallburg—L. W. N.

A. Filix-mas (L.) Sw. Male-fern.
> Rocky woods. Gilmer; near Glenville—V. M.

A. marginale (L.) Sw. M. & G.
> Rocky hillsides in rich woods. Gilmer; near Glenville—V. M. Kanawha; near Charleston—James. Fayette; near Nuttallburg—L. W. N. Grant; near Bayard.

A. acrostichoides (Michx.) Sw. Christmas Fern.
> Rocky woods. Upshur; beyond Buckhannon. Randolph; near Cricard. Gilmer; near Glenville—V. M. Fayette; near Nuttallburg—L. W. N.

Var. **Schweinitzii**, (Beck.) B. S. P. *Var. incisum*, Gray.
> Rocky woods. Jefferson; near Harper's Ferry—M. & G.

CYSTOPTERIS. Bernh.

C. bulbifera (L.), Bernh. Bladder-fern.
> Shaded ravines. Ohio; near Wheeling—M. & G. Fayette; near Nuttallburg, rare L. W. N.

C. fragilis (L.) Bernh. M. & G.
> Shady cliffs. Fayette; near Gauley Bridge along the Kanawha; near Nuttallburg—L. W. N.

ONOCLEA. L.

O. sensibilis. L. Sensitive Fern. M. & G.
> Moist meadows. Monongalia; the Flats. Gilmer; near Glenville V. M. Fayette—near Kanawha Falls—James; near Nuttallburg L. W. N. Randolph; near Valley Head. Upshur; near Buckhannon.

WOODSIA. R. Br.

W. obtusa (Spreng.) Torr M. & G.
> Rocks and cliffs. Fayette; near Nuttallburg L. W. N. Randolph; near Cricard.

DICKSONIA, L'Her.

D. punctilobula (Michx.), Gray. Dickson's Fern. *D. pilosius-cula*, Willd.
Moist shady places. Randolph: on Rich Mountains, alt. 1920 ft. Cheat mountains under Black Spruce, abundant. Gilmer: near Glenville—V. M. Fayette: near Nuttallburg, alt. 2400 ft.—L. W. N.; near Kanawha Falls and Loup Creek—James. Kanawha: near Charleston—James.

OSMUNDA. L.

O. regalis. L. Royal Fern. M. & G.
Swampy meadows. Upshur: near Randolph county line on Staunton Pike. Webster: Upper Glade. Gilmer: near Glenville—V. M. Fayette: near Nuttallburg—L. W. N. Preston: near Terra Alta and Cold Spring. Monongalia: near Camp Eden. McDowell: near Elkhorn.

O. Claytoniana. L. Clayton's Flowering Fern.
Low grounds. Preston: near Cranberry Summit—M. & G. Fayette: near Nuttallburg—L. W. N.

O. cinnamomea. L. Cinnamon Fern.
Marshy places. Fayette: near Nuttallburg, alt. 2000 ft.—L. W. N. Preston: near Cranberry Summit—M. & G. Randolph: along Shaver's Fork.

EQUISETACEÆ

EQUISETUM, L.

E. arvense. L. Field Horsetail. L. W. N., M. & G.
Moist, sandy fields. Frequent throughout the State.

E sylvaticum. L.
Damp rich woods. Mercer: near Ada. Monongalia: near Little Falls.

E. hyemale. L. Scouring Rush.
Wet wooded banks. Wirt: near Burning Springs.

E. lævigatum, Braun.
Clay banks along stream. Mercer: near Ingleside.

BRYOPHYTA.

SPHAGNA.

SPHAGNUM, L.

S. cymbifolium, Ehrh.
Common in wet glades, and in deep wooded rills.
Preston: Kingwood and Terra Alta. Monongalia: along
Tibb's Run. Webster: at Welsh, Long and Collett's Glades.
Fayette: glade above Nuttallburg. Randolph: in the Spruce
forests.

MUSCI.

(All the species in this class have passed through the
hands of Mrs. E. G. Britton, who has kindly deter-
mined them for this Flora.)

POLYTRICHACEÆ.

POLYTRICHUM, L.

P. commune. L.
Preston: on ground in open woods, Terra Alta.

P. Ohioense, Ren. & Card.   *P. formosum, Sull.* not Hedw.
Monongalia: on ground, Morgantown (1536); a large
form on ground Tibb's Run (1600); an extremely small form,
with minute capsules on sandstone boulder, loc. cit. (1611).
Mercer: on ground in oak woods, Bluefield (1453). Grant:
on decayed logs, Bayard. Fayette: near Nuttallburg—
L. W. N.

P. piliferum, Schreb.
Monongalia: on bare sandstone ledge, Falling Run.
(1299.)

P. tenue, Menz.   *Pogonatum brevicaule,* Beauv.
Monongalia: roadside banks, Morgantown; on ground,
Tibb's Run (1612). Fayette: near Nuttallburg—L. W. N.

CATHARINEA, Ehrh.

C. angustata, Brid.   *Atrichum angustatum,* Br. & Sch.
Monongalia: on ground in marshy spot, Morgantown

(1406). Mercer, rocks in rill, Beaver Spring (1495) Fay
ette: near Nuttallburg L. W. N.

C. undulata (L.), Web. & Mohr   *Atrichum undulatum*, Beauv.
Monongalia: on ground in marshy spot, Morgantown
(1404) McDowell: on roots in stream, Elkhorn 1522 Fayette: near Nuttallburg—L. W. N.

GEORGIACEÆ.

GEORGIA. Ehrh.

G. pellucida (L.), Rabenh.   *Tetraphis pellucida*, Hedw.
Monongalia: on sandstone boulder, Tibb's Run (1606,
1610, 1634).

FISSIDENTACEÆ

FISSIDENS. Hedw.

F. adiantoides (L.), Hedw.
Monongalia: on shaly rocks under cliff, Cassville
(1423).

F. decipiens, Schimp.
Fayette: near Nuttallburg—L. W. N.

MNIACEÆ.

ASTROPHYLLUM, Neck.

A sylvaticum, Lindb.   *Mnium cuspidatum*, Hedw.
Monongalia: on soil, Morgantown (1359); on dry
boulder, Cheat River 1397; on stone in swampy spot,
Dille's (1583). Fayette: near Nuttallburg—L. W. N.

A rostratum Schrad., Lindb.
Grant: on wet logs, Bayard. Monongalia: on decay-
ed wood, The Flats (1377 McDowell: on roots in rill,
Elkhorn (1523).

A. punctatum L. Lindb.   *Mnium punctatum* (L.) Hedw.
Fayette: near Nuttallburg L. W. N.

A. hornum L., Lindb.
Monongalia: on sand in rill, Tibb's Run (1604).
Fayette: near Nuttallburg – L. W. N.

SPHÆROCEPHALUS, Neck.

S. heterostichus(Brid.), Britt.m.   *Aulacomnion heterostichum*, Br.
& Sch.
Monongalia: On coal entrance to coal pit, Georgetown (1379); hanging from sandstone boulder, Tibb's Run
(1607), Camp Eden (1392); on rocky ledge,Cassville (1414).
McDowell: on sandy bank of rill, Elkhorn (1520). Fayette:
near Nuttallburg— L. W. N.

BARTRAMIACEÆ.

BARTRAMIA. Hedw.

B. pomiformis(L.), Hedw.
Mercer: on bole dead tree, Bluefield (1478). Monon-
galia: on sandstone boulder, deep woods, Tibb's Run (1609).
Fayette: near Nuttallburg—L. W. N.

Var. **crispa**(Sw.), Schimp.
Monongalia: on rock ledge, Cassville (1417, 1418).

PHILONOTIS, Brid.

P. fontana(L.), Brid.
Mercer: on sandstone ledge in rill, Beaver Spring(1561)

BRYACEÆ.

BRYUM, L.

B. bimum, Schreb.
Monongalia: on shale under ledge, Cassville (1424.)

B. argenteum, L.
Monongalia: fissures between bricks of walks, Morgantown (1335). Fayette: near Nuttallburg—L. W. N.

B. proliferum(L.), Sibth.   *B. roseum*, Schreb.
Mercer: on roots of oak, Bluefield (1449). McDowell:
on decayed wood, Elkhorn (1502).

LEPTOBRYUM, Wils.

L. pyriforme(L.), Wils.   *Bryum pyriforme*, Hedw.
Monongalia: on sandstone boulder, Tibb's Run (1616,
1633).

FUNARIACEÆ

FUNARIA, Schreb.

F. hygrometica(L.), Sibth.
Monongalia: in soil on sandstone boulder, Fibb's Run
(1615, 1617). Fayette: near Nuttallburg—L. W. N.

Var. **patula.** Br. & Sch.
Monongalia: on rocks lining a spring, the Flats
(1376); in cinders of an old camp fire, Camp Eden (1293).

F flavicans. Michx.
Monongalia: on damp sand in a "burning." Little
Falls (1277); loc. cit., Morgantown(1339.)

PHYSCOMITRIUM. Brid.

P. pyriforme(L.), Brid.
Monongalia: on top of soil of field that had been
ploughed and harrowed only eight days before, Morgantown
(1278); on ground marshy spot, Dille's (1463).

TORTULACEÆ.

LEERSIA. Hedw.

L. streptocarpa(Hedw.), Lindb.   *L. contorta*, Wolf. *Encalypta*
                                                    streptocarpa, Hedw.
Mercer: face dry limestone cliff, Beaver Spring (1552.

TORTULA, L.

T. muralis(L.), Hedw.   *Barbula muralis* (L.), Trin.
Mercer: on sandstone ledge, Beaver Spring (1553).

BARBULA, Hedw.

B. humilis. Hedw.   *B. caespitosa*, Schwaegr.
Mercer: roots of oak, Bluefield (1417). Fayette:
Nuttallburg—L. W. N.

B. tortuosa(L.), Web. & Mohr.
Monongalia: in sand under boulder, Camp Eden
(1395).

MOLLIA, Schrank.

M. viridula(L.), Lindb.   *Weisia viridula*, Hedw.
McDowell: on ground, open woods, Elkhorn (1496,
1497). Fayette: near Nuttallburg—L. W. N.

DICRANACEÆ.

LEUCOBRYUM. Hampe.

L. glaucum(L.). Schimp.

Monongalia: on ground in woods. The Flats (1399). Fayette: Nuttallburg—L. W. N.

DICRANODONTIUM, Br. & Sch.

D. Virginicus, Britt. m. Sp. nov.

Monongalia: on sandstone boulder along a woodland path. Tibb's Run (1635).

Plants bright glossy green, stems matted below by a red tomentum, leafy nearly to apex, denudate roughened above, with a few leaves at summit; leaves erect or secund, straight or curled and twisted, often 5 mm. long, narrowly subulate from a short, thick base, caducous ones with a long, slender, smooth point; persistent ones serrate, blade inflexed cells densely chlorophyllose, filled with oil globules, those of the basal angles, clear. Dioecious, the antheridia terminal in conspicuous heads, bracts brown at base, apex subulate, serrate; perichætial bracts 3–4mm. long, from a short base, suddenly subulate, dentate at apex; pedicels lateral by the growth of innovations, 1½–2cm. long, pale, glossy yellow, twisted in two directions, very slender, arcuate when young, becoming erect before capsules mature. Capsule cylindric, ribbed only at the mouth, 1½–2mm. long, beak straight or curved, shorter than the capsule, peristome bright red, not deep set, teeth split unequally to middle, striolate at base, pale and granulose above, annulus none, spores small, calyptra cucullate, 2mm. long, beaked, entire. Maturing in summer.

Differs from European specimens of *D. longirostre* collected by Seringe; in the longer, paler, more slender, scarcely arcuate pedicels, longer capsules, peristome not deep set, and teeth split only to the middle, more united than figured in the Bryologia Europea. Table 88. It may be distinguished from *Campylopus Virginicus*, also remarkable for its caducous leaves, by the longer, more slender subulate point, which is entire or minutely serrate and smooth on the back, by the thick base, with inflexed blades, and by the shape of the basal cells at the angles.

D. Millspaughi, Britt. m. Sp. nov. *Campylopus flexuosus*, Sull. not Brid.

Monongalia: on sandstone boulder deep woods. Tibb's Run (1596).

Plant slight yellowish green, silky, cæspitose; stems matted with rufous tomentum at base, 1–3cm. long, a few

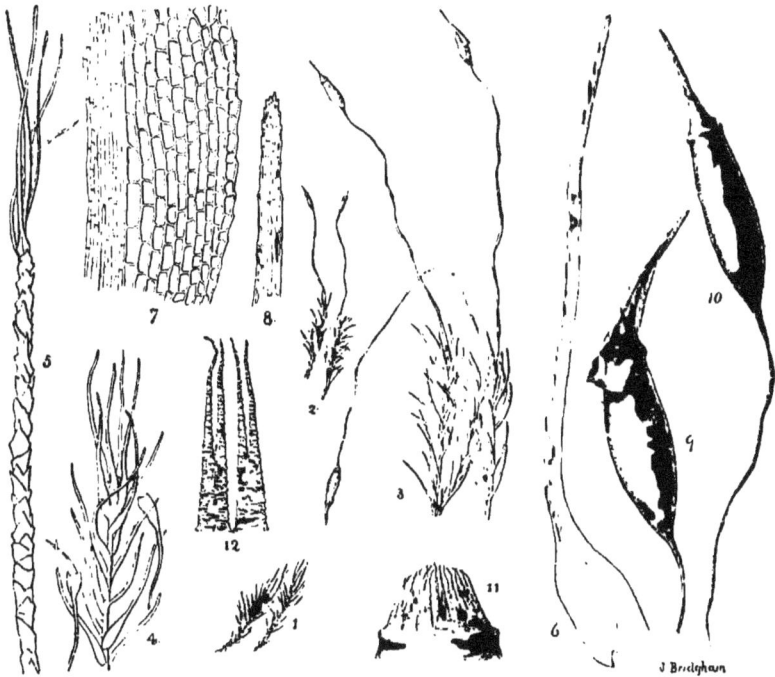

Dicranodontium Virginicum, n. sp.

J. Bridgham

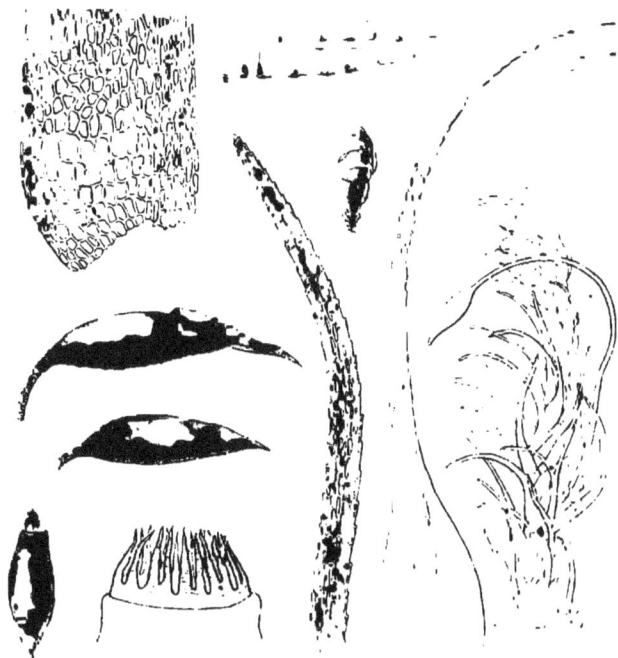

Dicranodontium Millspaughi, n. sp.

denudate, roughened by the fragments of the slightly caducous leaves. Leaves secund or erect-spreading, 15mm. long,
narrowly subulate from a broad base 1-1½mm. long, becoming tubular above with inrolled margins, basal angles not
auricled, filled by large hyaline cells to the base of the broad,
brown vein, those of the blade oblong or square next the
vein, becoming spindle-shaped and prosenchymatous toward
the margin, vein thick, excurrent into a dentate slender tip,
rough on back. Dioecious, perichaetium 5-7mm. long, bracts
sheathing half their length, tapering to a long, slender, ob-
scurely serrate tip, outer shorter, abruptly subulate, more
sharply serrate; pedicels recurved, burying the capsules
among the leaves, becoming erect when old, 5-8mm. long,
stout and twisted in two directions; capsules pyriform-cylin-
dric with a distinct neck, length about 1mm. without the lid
which is as long as the rest of the capsule, with a straight
beak, calyptra cucullate, entire; peristome red, connivent,
teeth deep set, slender, split to middle, or perforate to base,
striolate below, granulose above; annulus none, mouth bor-
dered by a dense, dark rim. Maturing in summer, old capsules persistent, not sulcate.

Differs from European specimens of *D. longirostre* in the
structure of the base of the leaf, lacking the suddenly inflated
basal auricles; differing also in the cells above the base, teeth
not split to base, occasionally only perforate. From *D. Virginicus* it may be distinguished by the less caducous leaves,
shorter, stouter, more arcuate pedicels, smaller capsules, and
longer sheathing perichaetium.

Through the kindness of Dr. Robinson I have been
able to compare these specimens with those collected by Sullivant on Grandfather Mt. in 1843. His also are fruiting,
and an excellent drawing is preserved, hence I am able to as-
sert that the specimens are almost identical, Sullivant's
showing no naked stems, but many of the leaves are caducous.
Dr. Braithwaite kindly compared the West Virginia specimens with *Campylopus pyriformis* sending me specimens of
this and the variety *Mulleri*, and sketches of the bases of the
leaves. It is evident that Sullivant was mistaken in refer-
ring his specimens to *C. flexuosus*, as they are more closely
allied to *Dicranodontium longirostre*, var. *alpinus*.

DICRANUM. Hedw.

D. flagellare, Hedw.

Monongalia: on decayed oak log, Tibb's Run (1593.)
Fayette: Nuttallburg—L. W. N.

D. scoparium, (L.). Hedw.

Grant: on ground in damp woods, Bayard. Monongalia: on ground, the Flats (1398); on decayed log, George-

town (1382); loc. cit., Tibb's Run (1601). Mercer: in tufts at base of stump, Bluefield (1476), and on decayed log (1464.) Fayette: near Nuttallburg— L. W. N.

D. fulvum. Hook.
    Fayette: Nuttallburg—L. W. N.

DICRANELLA, Schimp.

D. heteromalla L. s. Schimp.
    Monongalia: on ground under rail fence. The Flats (1362); on wet coal entrance of coal pit, Georgetown (1378); on ground in woods, Tibb's Run (1638). Fayette: Nuttallburg L. W. N.

DITRICHUM. TIMM.
(Lepotrichum, Hampe.)

D. pallidum (Schreb.), Hampe.   *Trichostomum pallidum*, Hedw.
    Mercer: on ground oak woods, Bluefield (1458). McDowell: on clay open woods, Elkhorn (1492, 1495, 1500.) Monongalia: on clay of open woods, Tibb's Run (1598). Fayette: near Nuttallburg—L. W. N.

CERATODON, Brid.

C. purpureus L.) Brid.
    Monongalia: on dry sandstone boulder, Morgantown (1390).; loc. cit., Tibb's Run (1633). Fayette: near Nuttallburg L. W. N.

GRIMMIACEÆ.

WEISSIA, Ehrh.

W. Americana (Palis.), Lindb.   *Ulota Hutchinsia*, Schimp.
    Monongalia: on dry sandstone boulder, Camp Eden 1390). Fayette: Nuttallburg—L. W. N.

ORTHOTRICHUM, Hedw.

O. Braunii. Br. & Sch.   *O. strangulatum*, Beauv.
    Monongalia: on bark living apple tree, Morgantown 1288.

HYPNACEÆ.

THUIDIUM. Br. & Sch.

T. recognitum Hedw.), Lindb.   *T. delicatulum*, Br. & Sch.
    Grant: on decayed logs, Bayard. Monongalia: on dry boulder, The Flats (1366); Tibb's Run (1608); on de-

cayed logs, Georgetown 1382. Mercer: on sandstone ledge,
Beaver Springs (1541): on box dead tree 1477: loc. cit.
Bluefield (1510). Fayette: near Nuttallburg L. W. N.

ANOMODON, Hook. & Tayl.

A. rostratus(Hedw.), Schimp.
Monongalia: on dry boulder, The Flats 1365: loc.
cit., Camp Eden (1391). Mercer: on sandstone ledge in rill,
Beaver Spring (1555): on hole living oak, Bluefield (1456,
1536). Fayette: Nuttallburg - L. W. N.

A. attenuatus Schreb.), Huebn.
Monongalia: on dry boulder, The Flats (1363, 1367),
Mercer: completely covering large limestone ledges in open
woods, Beaver Spring (1531).

AMBLYSTEGIUM, Br. & Sch.

A. adnatum, Hedw.
McDowell: on flat stone in deep woods, Elkhorn
(1498).

A. serpens(L.), Br. & Sch.
Monongalia: on wet rotten long, Granville (1298): on
twigs in rill, Dille's (1402). McDowell: on pebble in deep
woods, Elkhorn (1519).

Var. **orthocladon** Beauv.), Aust.
Monongalia: on rocks lining a spring, The Flats
(1375). Mercer: on wet limestone ledge, Beaver Spring 1358.

A. varium(Hedw.), Lindb.   *A. radicale*, Br. & Sch.
Monongalia: on rocks lining a spring, The Flats (1371):
on wet rocks in stream, Cassville (1421). Mercer: on limestone ridge, Beaver Spring (1535): on decayed log, Bluefield
(1488, 1536). McDowell: on top of stump in dark, deep
woods, Elkhorn (1521).

A. irriguum(Hook. & Wils.), Br. & Sch.
Mercer: on sandstone ledge in rill, Beaver Spring
(1556, 1559). Monongalia: on rocks in rill, Tibb's Run 1592).

A. riparium(L.), Br. & Sch.
Monongalia: on stone in stream, Falling Run (1331)
McDowell: loc. cit., Elkhorn (1512

A. chrysophyllum(Brid.), De Not. *Hypnum chrysophyllum*, Brid.
Monongalia: on old beech log, Morgantown (1405):
Mercer: on ground, Beaver Spring (1536).

HYPNUM, L.

H. denticulatum, L.
Monongalia: on walls of dark dripping limestone

cave, Cheat river: on stone in swampy place, Morgantown
(1405); on sandstone boulder, and on pebbles in stream.
Tibb's Run (1614.)

H. palustre, ?
Monongalia: on stone in marshy spot, Morgantown
(1584); on rocks under a fall. Casville (1422.)

H. molle. Dicks.
Monongalia: on stone in marshy place. Dille's (1584.)

H. proliferum. L. *H. splendens.* Hedw.
Randolph: in dense spruce forests, where it carpets
almost everything beneath the trees. Cheat Bridge. Grant:
notul. idem., Bayard.

H. rutabulum. L.
Monongalia: on rocks in rill. Tibb's Run (1591).

H. recurvans. Schwaegr.
Tucker: on decayed logs, etc., Blackwater Falls (990-2).
McDowell: loc. cit., Elkhorn (1499, 1507). Monongalia: on
bole of tree. Tibb's Run (1597).

H. microcarpum, C. Muell.
Monongalia: on bark of hemlock tree, Cheat River(1389).

H. hians, Hedw.
Mercer: on damp, decayed bark. Bluefield (1840).

H. demissum. Wils. *Rhynchostegium demissum,* Br. & Sch.
Monongalia: on stones in rill. Tibb's Run (1619); and
on wet rocks (1595).

BRACHYTHECIUM. Schimp.

B. salebrosum. Br. & Sch. *Hypnum plumosum.* Huds.; *H. salebrosum,* Hoffm.
Fayette: near Nuttallburg—L. W. N.

STEREODONTACEÆ.

THELIA. Sull,

T. hirtella Hedw., Sull.
McDowell: on bark living beech. Elkhorn (1493.)

T. asprella Schimp., Sull.
Mercer: on bole living Cornus florida. Beaver Spring
1535.

HYLOCOMIUM. Br. & Sch.

H parietinum L., Lindb. *Hypnum. Schreberi,* Willd.
Monongalia: on ground shade of Hemlocks. Laurel
Hill (1615.)

H. triquetrum L. i. Br. A Sch.

Monongalia: on _round of shad of hemlock. Laurel Hill (1649.)

CAMPYLIUM. Mitt.

C. hispidulum Brid. Mitt. *Hypnum hispidulum.* Brid.

Mercer: on ground, oak woods, Bluefield 1452.

C. chrysophyllum.

Mercer: on decayed log, damp place, Bluefield (1190, 1191).

Var. **tenellus**.

Mercer: on bole dead tree, Bluefield 1179); on log, damp place 1186.

PTILIUM. De Not.

P. crista-castrense L.). De Not. *Hypnum crista-castrensis.* L.

Plentiful on ground, rocks, logs, etc., in the dense spruce forests. Grant: near Bayard. Randolph: near Cheat Bridge.

STEREODON. Mitt.

S. imponens Hedw.). Brid. *Hypnum imponens.* Hedw.

Monongalia: on decayed log, Georgetown (1381); loc. cit., Tibb's Run (1602. Fayette: near Nuttallburg- L.W.N.

S. cupressiforme L.). Brid. *H. cupressiforme.* L.

Mercer: on damp decayed log, Bluefield (1487).

S. curvifolius Hedw.., Brid. *H. curvifolium.* Hedw.

Monongalia: on decaye l oaks, Little Fal's 1276); near Morgantown (1311); near Georgetown (1380); near Cassville (1420); on ground, Georgetown (1382). McDowell: on decayed log, Elkhorn 1517. Mercer: loc. cit., Bluefield 1485), and Beaver Spring 1491. Fayette: near Nuttallburg— L. W. N.

PYLAISIA, Br. & Sch.

P. velutina. Br. A Sch.

Monongalia: on bark living apple tree, Morgantown (1289.)

PLAGIOTHECIUM. Br. & Sch.

P. denticulatum. Br. & Sch.

Fayette: near Nuttallbur_ — L. W. N.

P. deticulatum. *var.* **densum.**

Monongalia: on sandstone boulder, Tibb's Run (1612.)

P. Sullivantiæ. Schimp.
Monongalia: on sandstone boulder in deep woods.
Tibb's Run (1618.)

CYLINDROTHECIUM, Br. & Sch.

C. seductrix Hedw., Sull.
Monongalia: on bark living apple tree, Morgantown
(1290); on bark in oak woods, Bluefield (1450). Fayette:
near Nuttallburg—L. W. N.

C. cladorhizans Hedw.), Schimp.
Mercer: on decayed log, damp place, Bluefield (1489).
Fayette: near Nuttallburg—L. W. N.

ENTODON, C. Muell.

E. palatinus (Neck.), Lindb.　　*Platygyrium repens*, Br. & Sch.
Monongalia: on decayed log, Tibb's Run (1603).
Fayette: near Nuttallburg—L. W. N.

NECKERACEÆ.

NECKERA. Hedw.

N. pennata (L.), Hedw.
Tucker: on tree trunks, Blackwater Falls (965).

CLIMACIUM. Web. & Mohr.

C. Americanum, Brid.
Monongalia: on moist sandstone ledge, Cassville(1413).

LEUCODON, Schwaeger.

L. julaceus (Hedw.), Sull.
Mercer: on limestone ledge, Beaver Springs (1532);
McDowell: on rocks in rill, Elkhorn (1508). Monongalia:
on oak log, Tibb's run (1590).

L. brachypus, Brid.
Grant: on wet rotten log, Bayard (937).

HEDWIGIA, Ehrh.

H. ciliata, Ehrh.
Monongalia: on dry, exposed boulders and rocks, The
Flats (1400).

HEPATICÆ.

(The following species are arranged in accordance with
Dr. A. W. Evans' "Arrangement of the Genera of
Hepaticæ;" Dr. Evans has kindly looked over and
identified all the numbers here reported. - C. F. M.

JUNGERMANNIACEÆ.

FRULLANIA, Raddi.

F. Asa-Grayana. Mont.
Monongalia : on sandstone boulder, Tibb's Run (1654),
Randolph: clinging to face of dry sandstone boulder, Pickens (2206).

JUBULA, Dumort.

J. Hutchinsiæ Hook . Dum. *var.* **Sullivantii,** Spruce.
McDowell : on rocks in stream, Elkhorn (1509). Monongalia : on sandstone boulder, Tibb's Run (1655).

LEJEUNEA, Libert.

L. clypeata (Schw.). Sull.
Monongalia : on sandstone boulder, Tibb's Run (1656.)

RADULA. Nees.

R. Xalapensis. Mont.
Mercer: face of limestone cliff, Beaver Spring (1551.
Agrees with Hep. Am. 104. Rare.

R. tenax, Lindb.
Monongalia : on sandstone boulder in deep woods,
Tibb's Run (1657.) Grant: on bark living cherry, deep
woods, Bayard (2060.)

PORELLA. Dill.

P. platyphylla L. . Lindb.
Monongalia : on bark living apple tree, Morgantown
(1291, 1292): on sandstone boulder, Tibb's Run 1658.
Mercer: on oak log, Bluefield (1418): on limestone ledge,
Beaver Spring (1530).

P. pinnata, L.
Fayette : on rocks in mist of falls, near Gauley Bridge
(607).

TRICHOCOLEA. Dumort.

T. tomentella (Ehrh.). Dum.
McDowell : growing with Catharinea undulata on roots
in rill, Elkhorn (1522 pt.). Grant : on wet sand in deep ravine, Bayard (2040.

HERBERTA. S. F. Gray.

H. adunca(Dicks.) S. F. G.

Monongalia: on sandstone boulder in deep woods, Tibb's Run. (1659.)

Other U. S. stations for this species are: Virginia, White Top Mt. —Mrs. Britton. North Carolina —Mr. Small. New York, Caatskills—Austin. New Jersey, Greenwood Mts.—Austin.

BAZZANIA. S. F. Gray.

B. trilobata(L.). S. F. G. *Mastigobrium trilobatum*, Nees.

Monongalia: on bole of tree in deep woods, Tibb's Run (1639, 1640): in wet depression sandstone boulder, loc. cit. (1660). Fayette: in deep woods—L. W. N. Grant: on wet Hemlock log, Bayard (2010.)

B. deflexa(Mart.), Underw.

Monongalia: on sandstone boulder in deep woods, Tibb's run (1661.)

CEPHALOZIA, Dum.

C. multiflora. Spruce.

Monongalia: on sandstone boulder and on ground in deep woods, Tibb's run (1662). Grant: on wet Hemlock log, deep woods, Bayard (2080).

C. curvifolia(Dicks.). Dum.

Monongalia: with the last, on sandstone boulder deep woods, Tibb's run (1663.) Grant: on wet dead bark deep woods, Bayard (2021).

ODONTOCHISMA. Dum.

O. Sphagni(Dicks.), Dum.

Monongalia: among mosses on sandstone boulder, deep woods, Tibb's Run(1664).

BLEPHAROSTOMA. Dum.

B. trichophyllum(L.), Dum.

Monongalia: on ground and sandstone boulder, deep woods, Tibb's Run (1665).

KANTIA, S. F. Gray.

K. Trichomanis(L.), S. F. G.

Tucker: on wet decayed logs, near Blackwater Falls

(993). Monongalia: on wet ground (1599), and on sandstone boulder in deep woods (1666), Tibb's Run. Randolph: on damp sand. Pickens (2207).

ANEURA, Dum.

A. multifida (L.), Dum.
Grant: on wet dead bark (3030), and wet decorticated wood (2070); in deep wooded ravine, Bayard.

GEOCALYX, Nees.

G. graveolens (Schrad.) Nees.
Monongalia: on ground and sandstone boulders, deep woods, Tibb's run (1667.)

SCAPANIA, Dum.

S. nemorosa(L.), Dum.
Monongalia: on damp sandstone boulder deep woods, Tibb's run (1668). Grant: on wet Hemlock log in deep woods, Bayard (2011). Randolph: on clay near a spring. Pickens, (2212).

DIPLOPHYLLUM. Dum.

D. taxifolium(Wahlenb.), Dum.
Monongalia: on sandstone boulder deep woods, Tibb's run (1669.)

PLAGIOCHILA, Dumort.

P. Virginica. Evans: sp. nov.
Growing in wide, depressed, and intricate tufts; stems ascending from a prostrate caudex, simple or sparingly branched, sometimes geniculate and rooting at the joints, otherwise eradiculose; leaves contiguous or somewhat imbricated, widely patent, ovate or rhomboid-ovate, the dorsal margin decurrent, slightly reflexed, entire, the ventral margin plane or reflexed at base, mostly entire, the apex broad, rounded or truncate, sharply and irregularly spinulose; amphigastria none

Stems 1 to 3 cm. long, with the leaves 1 to 2 mm. wide; leaves 1.2 mm. long, 0.7 mm. wide; spines short, acute, separated by rounded sinuses, varying in number from 2 to 3 on each leaf, usually 4 or 5; leaf-cells averaging 0.025 mm. in diameter in middle of leaf, thin-walled and scarcely thickened at the angles.

Mercer: on walls of dry limestone cave, Beaver Spring (1550).

Description of Figures
Fig. 1. Plants, natural size.
" 2. Apex of stem, dorsal view x 11.
3. Part of stem, ventral view x 11.
1. Apical teeth of leaf x 115.
5. Cels from middle of leaf x 225.

P. porelloides (Torr.), Lindb.
Monongalia: on sandstone boulders in moss, Tibb's Run (1700). Grant on wet stones (2000, 2050), and in wet sand (2041), in deep ravine, Bayard.

JUNGERMANNIA. Michx.

J. exsecta. Schmid.
Grant: on wet hemlock log in deep wooded ravine, Bayard (2012).

PELLIA, Raddi.

P. epiphylla (L.). Corda.
Randolph: on clay near a spring, Pickens (2211).

HARPANTHUS, Nees.

H. scutatus Web. & Mohr., Spruce.
Monongalia: on ground and sandstone boulder, deep woods, Tibb's Run (1670). Grant: on damp dead bark. Bayard (2020, 2031).

METZGERIA, Raddi.

M. conjugata, Lindb.   *M. furcata,* Dum. in. pt.
McDowell: on bark of beech, Elkhorn (1513). Monongalia: on bark of twig (1671).

MARCHANTIACEÆ.

MARCHANTIA. L.
M polymorpha. L.
Randolph: on ground in burnt place, summit of Point Mt. alt. 3700 ft. Grant: similar situation, near Bayard. Tucker: loc. cit., near Blackwater Falls. Gilmer: near Glenville V. M. Monongalia: among damp mosses base of sandstone boulder, Tibb's run: between bricks of sidewalk with Bryum argenteum, Morgantown.

CONOCEPHALUS. Necker.

C. conicus (L.). Dum.
Monongalia: on ground and sandstone boulders deep woods, Tibb's run (1672.)

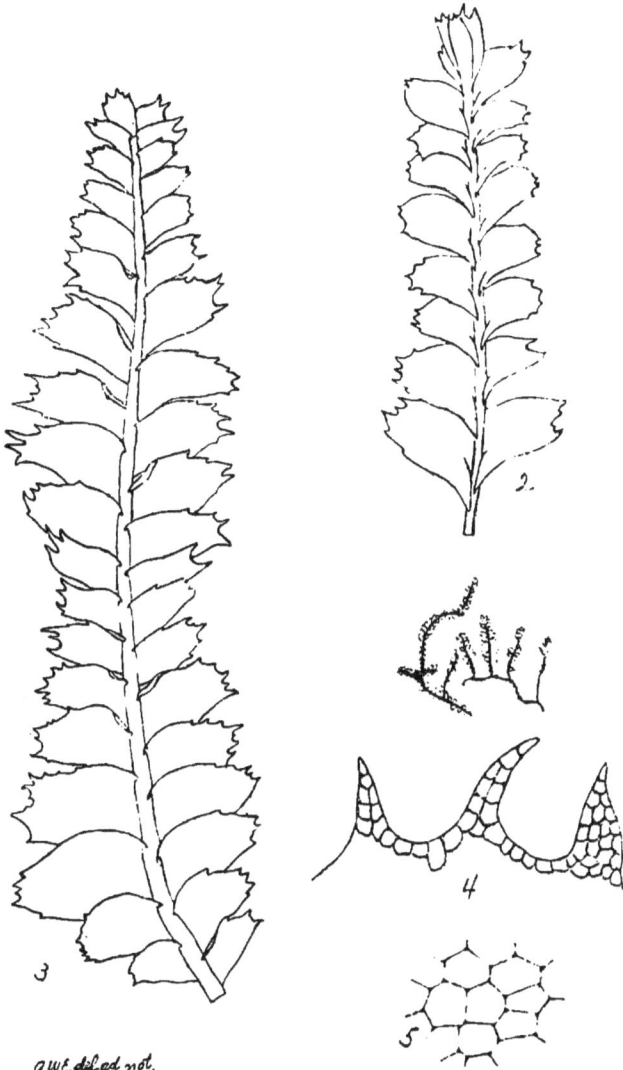

a.w.e. del ad nat.

Plagiochila Virginica, n. sp.

THALLOPHYTA.

LICHENES.

(The few specimens that I have gathered in this class, incidental to the collection of the higher orders, have been kindly examined by Dr. J. W. Eckfeldt.)

USNEÆ.

USNEA. Ach.

U barbata(L.) Fr.
Wirt: on old trees, common. Burning Springs. Randolph: on Rhododendron max. common. Cheat Bridge. Mercer: on oak twigs, Bluefield.; and elsewhere about State common on trees, rocks, and old fence rails.

Var. **florida**, Fr.
Mercer: on oak twigs and chips among dead leaves. Bluefield.

PARMELIEÆ

PARMELIA. Ach.

P. Borreri, *var.* **rudecta**. Tuck.
Monongalia: on bark living locust tree. Falling Run (1531).

P. caperata(L.) Ach.
Monongalia: on sandstone rocks and base of beech. Falling Run (1283).

P. olivacea(L.). Ach.
On Liriodendron log newly felled. Monongalia: Falling Run (1343).

PHYSCIA, DC.

P. lucomela(L.), Michx.
Mercer: with moss on wet limestone ledge, Beaver Spring (1539).

PELTIGERIEÆ.

STICTA. Schreb.

S. pulmonaria(L.), Ach.
  Wirt: on trunks of oaks, near Burning Springs (327).

S. herbacea(Huds.), Ach.
  McDowell: on oaks, Elkorn.

PELTIGERA, Willd.

P. aphthosa(K.), Hoffm.
  Monongalia: on rock ledge, near Cassville.

LECANOREÆ.

PLACODIUM, DC.

P. cerinum(Hedw.), Naeg. & Hepp.
  Monongalia: on bark of beech, Falling Run (1357).

LECANORA, Ach.

L. astra.(Huds.), Ach.
  Monongalia: on bark Liriodendron log, newly felled,
  Falling Run (1342); on flat exposed surface sandstone rock,
  same locality (1287).

CLADONIEÆ.

CLADONIA, Hoffm.

C. Mitrula, Tuck.
  Monongalia: old beech log, Falling Run (1346).

C. pyxidata(L.), Fr.
  Monongalia: along Falling Run, on bare sandstone
  rocks (1281); on moss in clay soil (1285); on decayed log
  (1338); base of beech in soil (1282).

C. gracilis(L.), Nyl.
  Monongalia: along Falling Run, on decayed log (1337);
  among mosses on clay soil (1286).

C. rangiferina L., Hoffm.
  Monongalia: on moss, Falling Run (1361).

C. furcata, *var.* **racemosa,** Floerk.

Monongalia : large patches on ground under chestnuts Dille's. Mercer : same growth under oaks, near Beaver Spring.

C. cristatella. Tuck.

Monongalia : on an old decayed log. Falling Run (1336).

DIATOMACEÆ

(Only one attempt has been made so far to collect species in this sub-class : this was a dredging in a small pasture pool near Morgantown. The result was kindly examined, and the species identified, by the Rev. Albert Monn, Jr.)

Cymbella gastroides, Kuetz.
    turgida(Grun.), Greg.

Stauroneis Phœnicentron, Ehrb.

Navicula viridis, Kuetz.
    major. Kuetz.
    nobilis(Ehrb.). Kuetz.
    rhomboides, Ehrb.
    borealis (Ehrb.), Kuetz.
    trinodis. Lewis.
    var. ——— ———

Achnanthes lanceolata, Breb.

Synedra Ulna(Nitzsch.), Ehrb.

Nitzschia Amphioxys, var. **intermedia,** Grun.

FUNGACEÆ.

AGARICACEÆ.

Leucosporea.

AMANITA. L.

A muscaria, L.

Rooted on buried birch limb. Grant: near Bayard.

Forma. ———

A wartless form, apparently of this species, occurs on leaf mold in deep woods. Grant: near Bayard.

COLLYBIA, Fr.

C. dryophila. Bull.

In moss on log, in deep woods. Grant: near Bayard.

C. radicata. Rehl.

On leaf mold. Monongalia: Rich woods near Morgantown.

MYCENA. Fr.

M. galericulata. Scop.

Under dead oak twig. Under bark oak log. Monongalia: rich woods near Morgantown.

OMPHALLA, Fr.

O. campanella, Batsch. (in part).

On leaf mold, base of chestnut. Preston: near Terra Alta.

MARASMIUS. Fr.

M. opacus. B. & C.

On dead branch Rhododendron maximum, common. Grant: near Bayard.

M. rotula (Scop.). Fr.

On dead birch limb. Grant: near Bayard.

LENTINUS. Fr.

L. strigosus. Schw.

On dead Birch log. Grant: near Bayard.

PANUS. Fr.

P. stipticus (Bull.), Fr.
Under dead branch Oak. Monongalia: near Morgantown.

LENZITES. Fr.

L. sepiaria(Wulf.), Fr.
On decorticated spruce stumps. Tucker: near the Falls of Blackwater.

Forma —— ——
A resupinate form. On dead hemlock logs. Grant: near Bayard.

SCHIZOPHYLLUM, Fr.

S. commune, Fr.
On bark dead oak. Grant: near Bayard. On dead apple twig. Monongalia: near Morgantown.

Rhodospora or *Hyporrhodia*.

VOLVARIA. Fr.

V. bombycina(Pers.), Fr.
Monongalia: on insect (?), Morgantown.

Melanospora.

STROPHARIA. Fr.

S. stercoraria. Fr.
On decaying vegetable matter. Preston: near Terra Alta.

HYPHOLOMA. Fr.

H. sublateritium. Schaeff.
Under bark ash log. Monongalia: near Morgantown.

PANÆOLUS, Fr.

P campanulatus, L.
On cow droppings, in deep coniferous woods. Grant: near Bayard.

POLYPORACEÆ.

POLYPORUS, Fr.

P. abietinus(Dicks.). Fr.
    On fallen Hemlock logs. Grant: near Bayard.

P. adusts(Willd.). Fr.
    On dead Sumach branch. In decayed hickory stump.
Monongalia: near Morgantown.

P. applanatus (Pers.). Fr.
    On dead Sugar Maple and Oaks. Wood: near Kanawha Station. McDowell: near Elkhorn. On dead ash log.
Grant: near Bayard. Monongalia: near Morgantown. Frequent throughout the State.

P. Berkeleyi. Fr.
    In dry exposed hollow in Oak stump. Monongalia:
near Morgantown.

P. carneus. Nees.
    On dead and decorticated Spruce stumps. Tucker:
near the Falls of Blackwater.

P. fomentarius L.) Fr.
    On dead Birch. Grant: near Bayard.

P. hirsutus(Wulf.). Fr.
    On dead birch and apple twigs. On bark Liriodendron log. Monongalia: near Morgantown.

Forma ————
    A small form, with white spores. On roots of fallen
birch. Grant: near Bayard.

P. lucidus Leyss.). Fr.
    On dead hemlock logs. Preston: near Terra Alta.

P. sulphureus. Fr.
    On decaying oak stump. Monongalia: near Morgantown.

P. umbellatus. Fr.
    In dry exposed hollow of an oak stump. Monongalia: near Morgantown.

P. versicolor(L.). Fr.
    On decorticated Spruce stumps. Tucker: near the
Falls of Blackwater.

FOMES. Fr.

F. rimosus, Berk.
Exposed on dead Yellow Locust log. Monongalia; near Morgantown.

POLYSTICTUS. Fr.

P. hirsutus. Fr.
On dead Apple twig. Monongalia: near Morgantown.

P. pergamenus, Fr.
On dry exposed Oak railroad tie. Monongalia: near Morgantown.

P. cinnabarinus(Jacq.). Fr.
Monongalia: on dead limbs of cherry. Morgantown.

P. versicolor, Fr.
Under bark oak log. Monongalia: near Morgantown.

TRAMETES, Fr.

T. sepium, Berk.
On dry exposed oak railroad ties. Monongalia: near Morgantown.

MYRIADOPORUS. Pk.

M. induratus, Pk.
Probably only an imperfect condition of Poria abducens, Pers *fide* Pk. Top of decayed oak stump. Monongalia: near Morgantown.

GLŒOPORUS, Mont.

G. conchoides. Mont.
On oak chips. Monongalia: near Morgantown.

MERULIUS, Hall.

M. tremellosus, Schrad.
Under bark oak log. Monongalia: near Morgantown.

HYDNACEÆ.

IRPEX. Fr.

I. lacteus. Fr.
On dead sumach. Monongalia: near Morgantown.

THELEPHORACEÆ.

STEREUM, Fr.

S. complicatum, Fr.
On roots of fallen birch. Grant: near Bayard. On dry oak railroad ties. Monongalia: near Morgantown.

S. frustulosum (Pers.), Fr.
On oak log. Monongalia: near Morgantown.

S. sericeum, Schw.
On dead birch twig. Grant: near Bayard.

S. sulphuratum, B. & Rav.
On White Oak log. Monongalia: near Morgantown.

S. versicolor, Fr., *var.* ——
On dry Oak railroad ties. Monongalia: near Morgantown.

HYMENOCHÆTE. Lev.

H corrugata (Fr.), Lev.
On decorticated birch limb. Grant: near Bayard.

EXOBASIDIUM, Weron.

E. Rhododendri, Cram.
Forming "cups" near the tips or margins of living leaves of Rhododendron maximum. Common in Grant and Tucker counties. I understand from Prof. Peck. that this is his first knowledge of the occurring of this species in North America.

CLAVARIACEÆ.

CLAVARIA, Vaill.

C. flaccida, Fr.
On leaf mold, in deep woods. Grant: near Bayard.

TREMELLACEÆ.

HYPSILOPHORA, Cke.

H fragiformis (Nees), Cke.
On bark dead Beech. Grant: near Bayard.

LYCOPERDACEÆ.

BOVISTA, Pers.

B. pila. B. & C.

Monongalia: free on open ground, Morgantown.

LYCOPERDON, Tourn.

L. pyriforme. Schæff.

Under bark oak log. Monongalia: near Morgantown, near Little Falls.

SCLERODERMA. Pers.

S. vulgare. Fr.

On spruce chips. Tucker: near Falls of Blackwater. Monongalia: plentiful on clay of woodland path; Tibb's Run. Grant: on dead logs, Otter Fork of Cheat.

UREDINACEÆ.

Amerospora.

UROMYCES. Lev.

U. appendiculata (Pers.), Lev.

On living leaves of Pole Bean. Monongalia: near Morgantown.

U. Hedysari-paniculati (Schw.), Farl.

On living leaves *Desmodium canescens*. Mason: near Point Pleasant.

U. Lespedezæ (Sch.), Pk.

On living leaves *Lespedeza violacea*. Monongalia: near Morgantown.

U. Trifolii (A. & S.), Winter.

On living leaves *Trifolium pratense*. Mason: near Point Pleasant.

Didymospora.

PUCCINIA, Pers.

P. flosculosorum (A. & S.), Rohl.

On living leaves *Cnicus lanceolatus*. Mason: near Point Pleasant.

P. Pimpinellæ St.). Link.
On living leaves *Osmorrhiza Claytoni.* Monongalia: near Morgantown.

P. Rubigo-vera
On living leaves *Triticum sativum.* Wood: near Kanawha Station.

P. sauveolens Pers.), Pk.
On living leaves *Cnicus lanceolatus.* Wood: near Kanawha Station.

GYMNOSPORANGIUM, DC.

G. macropus. Link.
Mercer: plentiful on living Junipers, near Princeton.

Phragmosporae.

PHRAGMIDIUM, Lk.

P. Potentillae Pers.), Karst.
Monongalia: uredo on living leaves *Potentilla Canadensis,* near Morgantown.

COLEOSPORIUM. Lev.

C. Sonchi-arvensis(Pers.) Lev.
On living leaves of *Veronia Noveboracense,* common. Mason: near Point Pleasant.

C. Senecionis(Pers.). *Peridermium Pini,* Auth.. *P. acicola,* Rabh.
Aecidium Pini, Pers.
Wood: aecidium on living leaves *Pinus mitis,* near Lockhart's Run.

Uredinea Inferiores.

ÆCIDIUM. Pers.

A. Houstonianum. Schw.
Monongalia: spermagonia on living leaves *Houstonia caerulea,* near Morgantown.

ROESTELIA, Rebent.

R. Pyratum. Schw.
On living leaves roan-beauty apple. Wood: near Lockhart's Run. Cabell: near Huntington.

PERIDERMIUM. Chev.

P. Peckii.
Pocahontas: on living leaves *Tsuga Canadensis*, near Traveler's Repose.

P. balsameum. Pk.
Under surface living leaves *Abies Balsamea*. Randolph: at Shades-of-Death.

CÆOMA. Link.

C. nitens. Schw.
On living leaves *Rubus hispidus*. Monongalia: near Morgantown.

USTILAGINACEÆ.

Amerospora.

USTILAGO, Pers.

U. Maydis (DC.), Cda.
On living ears and tassels of sweet corn. Monongalia: near Morgantown, prevalent (1891).

U. segetum (Bull.), Ditm.
On living heads of wheat and oats. Monongalia: near Morgantown. Lewis: near Alum Bridge. Taylor: near Thornton.

U. Tritici (Pers.), Jensen.,
On living leaves of wheat. Monongalia: near Morgantown.

Dictyospora.

UROCYSTIS, Rabenh.

U. Anemones, Schroeb.
Monongalia: on living leaves and under bark of *Jetæa*. Morgantown.

PERONOSPORACEÆ.

PHYTOPHTHORA. D. By.

P. infestans Mont. , D. By.
On living potato leaves and tubers. Monongalia: near Morgantown.

P. obovata, Bonor.
  On ? Tucker: near Davis.

CYSTOPUS. DeBy.

C. candidus(Pers.), Lev.
  Monongalia: on living leaves *Dentaria diphylla,* Little
Falls.

PERONOSPORA. Cda.

P. obovata, Bon.
  Preston: on living leaves *Spergula arv.,* Terra Alta.

P. viticola.
  On Grapes. Monongalia: near Morgantown.

ENTOMOPHTHORACEÆ.

EMPUSA, Cohn

E. Grylli(Fr.). Nowakow.
  On tufted catapilar, on locusts, and on *Musca.* Monongalia: near Morgantown, plentiful (1891).

E. Muscæ(Fr.). Cohn.
  On common house-fly: prevalent on a Tachina sp. on
maples in 1892: Monongalia: Morgantown.

PERISPORIACEÆ.

(*Erysiphea.*)
Amerospora.

PODOSPHÆRA. Kunze.

P. oxycantha(DC.). D. By.
  On living Cherry, Thorn, and Persimmon. Monongalia: near Morgantown.

P. triadactyla(Waller.). D. By.
  On living leaves cultivated Cherry. Cabell: near
Huntington.

SPHÆROTHECA, Lev.

S. Humuli(DC.), Burrill.
  On living leaves *Agrimonia Eupatoria.* Preston: near
Terra Alta.

UNCINULA, Lev

U. Ampelopsidis, Pk.

On grapes. Monongalia; near Morgantown

ERYSIPHE Hedw., DC

E. graminis. DC.

On living leaves *Poa pratensis*. Preston near Terra Alta.

SPHÆRIACEÆ

Phæosporæ.

HYPOXYLON, Bull

H. fuscum(Pers.). Fr.

On dead and decorticated maple. Grant near Bayard.

DOTHIDEACEÆ.

Hyalosporæ.

PHYLLACHORA, Fckl.

P. graminis(Pers.). Fckl.

On living leaves of *Aspirella Hysterix*. Fayette; near Nuttallburg.

PLOWRIGHTIA, Sacc.

P. morbosa(Schw.). Sacc.

Monongalia; on limbs of plum and cherry. Morgantown.

HYSTERIACEÆ.

LOPHIODERMIUM, Chev

L. Pinastri(Schrad.). Chev.

Hampshire; on living leaves *Pinus inops*. Romney

PEZIZACEÆ.

Hyalosporæ.

PEZIZA, L.

(*Mollisia Karst.*)

P. cinerea. Batsch.

Monongalia; on decayed log. Morgantown.

P. aurelia, Pers.

Monongalia; on dead leaf in rotten log. Little Falls.

P. scutellata, L.

On dead bark in water. Grant: near Bayard.

LACHNEA. Fr.

L. erinaceus(Schw.)

Under side Oak log. Monongalia: near Morgantown.

L. scutellata(L..)

On rotting Beech log, and under bark wet Oak log. Monongalia: near Morgantown.

BULGARIACEÆ.

Phragmospora.

CORYNE, Tul.

C. urnalis(Nyl.). Sacc.   *C. purpurea,* Fckl.
        Rotting beech log, and under bark oak log. Monongalia: near Morgantown.

GYMNOASCACEÆ.

(*Ejootseea.*)

TAPHRINA, Tul.

T. Pruni. Fckl.

On plums. Monongalia: near Morgantown.

T. deformans(Berk.), Tul.

On peach leaves. Jefferson: near Charlestown.

SACCHAROMYCETACEÆ.

SACCHAROMYCES, Myen.

S. cerevisiæ, Meyen.

In Pasteur's liquid left uncorked in laboratory.

S Mycoderma. Reess.

On same liquid as above.

SCHIZOMYCETACEÆ.

Bacubigea.

BACILLUS, Cohn

B. acidi-lactici.
In sour milk.

B. subtilis. Cohn.
In infusion of hay.   On boiled potato.

B. tuberculosss. Koch.
In sputa of consumptives.

B. ulna, Cohn.
On boiled white-of-egg.

B. ———
Isolated from blood of self when suffering from "La Grippe."

SPIRILLUM. Ehrb.

S. undula(Muell.), Ehrb.
In infusion of hay.

BACTERIUM, Ehr. et. Trev.

B. lineola, Cohn.
In infusion of radish.

B. termo. Dujard.
In various decomposing organic substance.

B. ——— sp. nov. Mss.
Isolated from dead locusts.

MICROCOCCUS (Hall.). Cohn.

M. amylovorus, Burrill.
In "fire-blight" of Pears.

M aurantiacus. Cohn.
Caught on sterilized potato.   In laboratory.

M. crepusculum, Cohn.
Found associated with *Bacterium termo* in decomposing infusions.

M. luteus, Cohn.
Caught on sterilized potato.   In laboratory.

M septicus. Cohn.
Found in blood of dead calf.

M. ureæ. Cohn.
In decomposing urine.

MYXOMYCETACEÆ
Tricophora.

TRICHIA. Hall.

T. chrysosperma Bull., DC.
Monongalia: on decayed wood, near Morgantown.

HEMIARCYRIA, Rostf.

H. clavata, Rostf.
Under bark of wet oak log. Monongalia: near Morgantown.

H. rubiformis, Rostf.
Under bark ash log. Under bark oak log. Monongalia: near Morgantown.

SPHÆRIOIDACEÆ.
Hyalosporæ.

PHYLLOSTICTA, Pers.

P. acericola, C. & E.
On living leaves *Acer saccharinum,* L. Putnam: near Buffalo.

P. Asiminæ, E. & Kell.
On living leaves of *Asimina triloba.* Monongalia: near Camp Eden.

P. Labruscæ. Thum.
On living leaves Concord Grape. Wood: near Lockhart's Run.

Scolecospora.

SEPTORIA. Fr.

S. Rubi. West. et B. & C
Wood: on living leaves *Rubus Canadensis,* Lockhart's Run.

S. Verbenæ. Ger.
On living leaves *Verbena urticifolia.* Jefferson: near Shenandoah Je.

S. Kalmiæcola(Sch.) B. & C.
Monongalia: on living leaves of *Kalmia latifolia*, Camp Eden.

LEPTOSTROMACEÆ

Phragmospora.

DISCOSIA, Fr.

D. maculæcola. Ger.
On living leaves *Disporum lanuginosum.* Grant: near Bayard.

ENTOMOSPORIUM, Lev

E. maculatum Lev.
On living leaves and fruit of the pear. Monongalia: near Morgantown.

MELANCONIACEÆ

Hyalosporæ.

COLLETOTRICHUM, Cda.

C Lindemuthianum. Scrib.
Pods of wax butter-bean. Monongalia: near Morgantown.

Scoleco-allantosporæ.

LIBERTELLA, Desm,

L. faginea, Desm. (*Quaternaria Personii*, Tul. *forma.*)
On bark of dead beech. Grant: near Bayard.

MUCEDINACEÆ.

Amerospora.

MONILIA, Pers.

M. fructigena, Pers.
On cherries. Monongalia: near Morgantown.

OIDIUM, Link.

O. monilioides. Lk. *Conidia of Erysiphe graminis,* DC.
Preston: on living leaves *Poa pratensis,* near Terra Alta.

O. leucoconium. Desm. *Conidia of Sphærotheca pannosa,* Lev.
Cabell: on living leaves of *Rosa* cult., near Huntington.

HYPHODERMA, Fr.

H. Dezmanzieri, Duly.
Wood; on living leaves *Pinus mitis*, Lockhart's Run.

DEMATIACEÆ.

Didymospora.

FUSICLADIUM, Bon.

F. dendriticum.(Wallr.), Fckl.
On living leaves and fruit of the Apple. Monongalia:
near Morgantown.

Phragmospora.

CERCOSPORA. Fr.

C. caulophylli, Pk.
On living leaves *Caulophyllum thalictroides*. Grant:
near Bayard.

C. smilacis. Thun.
On living leaves Smilax. Monongalia: near Camp
Eden.

Dictyospora.

MACROSPORIUM, Fr.

M. Tomato, Cook.
On cultivated sp. tomato. Monongalia: near Morgan-
gantown, not prevalent (1891).

SEPTOSPORIUM. Cda.

S. Equiseti, Pk. sp. nov. (MSS.).
Tips of living leaves of *Equisetum arvense*. Doddridge:
near Center Point. Monongalia: on campus.

TUBERCULARIACEÆ.

(*Tubercularica Mucedinea.*)
Amerospora.

TUBERCULARIA. Tode.

T. vulgaris
On dead Sumac limbs. Monongalia: near Morgantown.

CYLINDROCOLA. Bon.

C. dendroctoni. Pk. n. sp.
Sporodochia minute, forming irregular masses, soft,

somewhat waxy, white or whitish; sporophores slender, abundantly branched above, often compacted below into a short stem-like base, spores catenulate, short cylindrical, sub-truncate, colorless, .00016 to .0002 in. long, .0008 to .0001 broad.

On dead insects, *Dendroctonus frontalis*, beneath the bark of pine. Hampshire: near Romney, W. Va.

The insects are probably killed by this fungus as they lie dead in their burrows in the inner bark of the tree (*Pinus inops.*)

On some of the insects there is a cottony or flocculent mass of white mycelium interwoven in a somewhat reticulate manner, and collected in strings or bundles. It bears no fruit but is probably a luxuriant growth of the mycelium of this fungus.

Occasionally the fungus seems to spread from the insect to bark immediately adjacent to it.

Phragmospora.

BACTRIDIUM. Kunze.

B. flavum. K. & S.

Under bark wet Oak log. Monongalia: near Morgantown.

FUSARIUM. Desm.

F. Solani. Mart.

Found associated with the black rot on tomatoes that have fallen badly affected with the disease.

F. culmorum. Smith.

Monongalia: on heads living ripe wheat, Laurel Point.

"This specimen combines the characters of a number of so-called species, making it difficult to say which it really is. Probably they are all forms of one species." Prof. Peck (in letter).

Tubercularia Denutica.
Ameriospora.

EPICOCCUM, Lk.

E. neglectum. Desm.

On living leaves *Acana sativa* and *Catalpa Bignonioides.* Monongalia: near Morgantown.

CONCLUSION.

Although I have worked only two seasons among the plants of the State, as a side issue from my duties at the Experiment Station, I can not feel notwithstanding the assistance of those who have kindly contributed toward this catalogue that much more than a beginning has been made toward a knowledge of the plant life within our boundaries. However, this rich field already makes a good showing, even when compared with the almost complete work, done by many observers combined, in other states, as will be noted by an examination of the following table:

	W. Va.	Col.	Mass.	Ohio.	Ark.	Neb.	N. Y.	Ills.	N.J.
Genera	514	130	113	153	562	562	533	551	
Species*	1365	1115	1162	1232	1233	1258	1330	1431	1995

* Anthophyta and Pteridophyta only.

Summary of the Flora.

	Genera.	Species.	Varieties.	Forms.	Total.
Anthophyta.	504	1189	109	23	1321
Pteridophyta.	15	39	1	1	44
Bryophyta.	66	107	6		113
Thallophyta.	94	164		3	167
Total.	679	1499	119	27	1645

Of these there are native of the State, 1452
Foreign, 193

Total species, varieties, and forms, **1645**

—— O——

SUPPLEMENT.

FOSSIL FLORA.

In this preliminary list of the rich fossil flora of the State I give simply an alphabetical arrangement of the genera and species, not deeming it necessary to enter into the classification or more detailed designation of the geological and geographical distribution than a simple mention of the localities, on account of the undeveloped condition of the work in this branch.

The major part of this list is kindly contributed to this publication by Mr. R. D. Lacoe from his extensive library and collections, the latter containing a large amount of material as yet unworked.

The study of the palaeobotany of this State has been already ably begun by Profs. I. C. White, and W. M. Fontaine, (F. & W.), who have published a work entitled "The Permian Flora," issued as Vol. PP of the Pennsylvania, Second Geological Survey.

ALETHOPTERIS.

A. aquilina(Schl.),Schp.	Lower Barrens.	Ohio: Wheeling.
A. gigas,Gein.	Upper Barrens.	Marshall: Bellton.
A. grandifolia, Newby.	Conglomerate.	Fayette: Quinnimont
A. Helena,Lx.	Conglomerate.	Fayette: Quinnimont.
A. lonchitica,(Br.),Schp.	Conglomerate.	Fayette: Quinnimont.
A. Virginiana. F. & W	Waynesburg C.	Monongalia: Cassville.

ANNULARIA.

A. carinata,Guth	Waynesburg C.	Monongalia: Cassville. Doddridge: W.Union.
	Upper Barrens	Marshall: Bellton.
A. longifolia,Br.	Waynesburg C.	Monongalia. Cassville.
	Waynesburg C.	Doddridge: W Union.
A. minuta,Br.	Upper Barrens.	Monongalia: Blacksville.
A. radiata(Br.),St.	Waynesburg C.	Monongalia: Cassville.
A. sphenophylloides,Zenk,	Waynesburg C.	Monongalia: Cassville.
	Lower Barrens.	Ohio: Wheeling.
Var. intermedia,Lx.	Upper Barrens.	Monongalia: south of Jollytown, Pa.
	Upper Barrens.	Monongalia: Brown's Mills.

ARCHAEOPTERIS.
.A. obtusa, Lx.   Vespertine. Greenbrier: Lewis Tunnel.
.A. Alleghanensis(Meek.), Lx. Vespertine. Greenbrier: Lewis Tunnel.
.A. Roehschiana(Goep.), Lx. Vespertine. Greenbrier: Lewis Tunnel.
.A. Hibernica.Forb (Lx.). Vespertine. Greenbrier: Lewis Tunhel.

ASTEROPHYLLITES.
.A. aricularis, Daws.   Conglomerate. Fayette: Quinnimont,
                                                Sewell.
.A. rigidus, Br.   L. Productive C. M.
.A. sublaris, Lx.   L. Productive C. M. Kanawha:

BAIERA.
B. Virginiana, F. & W.   Upper Barrens.

CALAMITES.
C. cannaformis, Schloth. Lower Barrens. Ohio: Wheeling.
                            Conglomerate. Fayette: Quinnimont,
                                                  Sewell.
C. Cisti, Br.   Waynesburg C. Monongalia: Cassville.
                            L. Productive C. M. Boone: Short Creek.
C. Suckowii, Br.   Waynesburg C. Monongalia: Cassville.
                            Waynesburg C. Doddridge: W. Union.

CALLIPTERIDIUM.
C. Dawsonianum, F. & W. Waynesburg C. Doddridge: W. Union.
                            Waynesburg C. Monongalia: Cassville.
                                      Dent's Run, Georgetown.
C. grandifolium, F. & W. Waynesburg C. Doddridge: W. Union.
                          Waynesburg C. Monongalia: Dent's Run.
                                      Georgetown.
C. Grandini.(Br.), Lx. L. Productive C. M. Kanawha: Camp-
                                      bell's Creek.
C. Massilionum, Lx.   L. Productive C. M. Boone: Short Creek.
C. oblongifolium, F. & W. Waynesburg C. Monongalia: Cassville.
                                    Dent's Run. Georgetown.
                          Upper Barrens. Marshall: Bellton.
C. odontopteroides, F. & W. Waynesburg C. Monongalia: Arnetts-
                                      ville. Cassville. Dent's
                                      Run, Georgetown.
C. unitum, F & W   Upper Barrens.

CALLIPTERIS.
C. conferta, Br.   Washington C. Monongalia: Brown's
                                    Bridge, Brown's Mills
                                    Dunkard Creek

CARDIOCARPUS.
C. longicollis, Lx.   L. Productive C. M. Kanawha: Camp-
                                      bell's Creek.
C. Ollipticus var. rugosus, MSS. L. Productive C. M. Kanawha: Malden
                                      mine.

CARDIOPTERIS.
  C. frondosa. Schimp.   Vespertine   Greenbrier Lewis Tunnel.

CARPOLITHES.
  C. bicarpa, F.& W.   Waynesburg C.   Monongalia: Cassville.
  C. marginatus, F.& W.   Waynesburg C.   Monongolia: Cassville.

CAULOPTERIS.
  C. elliptica, F.& W.   Waynesburg C.   Monongalia: Cassville.
  C. gigantea, F.& W.   Waynesburg C.   Monongalia: Cassville.

CORDIATES.
  C. borassifolius (St.). Ung.   Lower Barrens.   Ohio: Wheeling.
  C. crassinervis, F.& W.   Waynesburg C.   Monongalia: Cassville.

CYMOGLOSSA.
  C. obtusifolia, F.& W.   Waynesburg C.   Monongalia: Cassville.
  C. brevilobata, F.& W   Waynesburg C.   Monongalia: Cassville.
  C. formosa, F.& W.   Waynesburg C.   Monongalia: Cassville.
  C. lobata, F.& W.   Waynesburg C.   Monongalia: Cassville.

EQUISITITES.
  E. elongatus, F.& W.   Waynesburg C.   Doddridge: W. Union.
  E. rugosus, Schimp.   Waynesburg C.   Doddridge: W. Union.
  E. striatus, F.& W.   Waynesburg C.   Doddridge: W. Union.

EREMOPTERIS.
  E. artemisiafolia (Br.), Schp.   L. Productive C.M.   Kanawha: Coalmont.
  E. Cheathami, Lx.   L. Productive C.M.   Kanawha: Bl'ksburg.

GONIOPTERIS.
  G. elliptica, F.& W.   Waynesburg C.   Monongalia: Cassville.
  G. Newberriana, F.& W.   Waynesburg C.   Doddridge: W. Union.
  G. oblonga, F.& W.   Waynesburg C.   Doddridge: W. Union.

GUILIELMITES.
  G. orbicularis, F.& W.   Waynesburg C.   Monongalia: Cassville.

LEPIDODENDRON.
  L. clypeatum, Lx.   L. Productive C. M.
  L. selaginoides, Sternb.   Conglomerate.   Fayette: Quinnimont.
                                        Sewell.
  L. Sternbergi, Br.   L. Productive C. M.
                   Vespertine.   Greenbrier: Lewis Tunnel.
  L. Veltheimianum, Sternb.   Vespertine.   Greenbrier: Lewis Tunnel.

LEPIDOSTROBUS.
  L. ornatus, L.& H.   L. Productive C. M.   Kittanning Coal.
  L. Salisburgi, Lx.   L. Productive C. M.   Coal river.

LESCUROPTERIS.
L. adiantites, Lx.
L. Moorei (Lx), Schp.

	Waynesburg C. Monongalia: Georgetown.
	Lower Barrens. Ohio: Wheeling.

MEGALOPTERIS.
M. Hartii, Andr.

Conglomerate.	Fayette: Quinnimont.
	Sewell.

M. Sewellensis, Font.

Conglomerate.	Fayette: Sewell.

NEMATOPHYLLUM.
N. angustum, F. & W.

Waynesburg C.	Monongalia: Cassville.
Waynesburg C.	Doddridge: W. Union.

NEUROPTERIS.
N. acutifolia, Br.

Lower Barrens	Ohio: Wheeling.
L. Productive C.	M. Upper Freeport C.

N. auriculata, Br.

Waynesburg C.	Monongalia: Cassville.
	Arnettsville.
Waynesburg C.	Doddridge: W. Union.
Upper Barrens.	Marshall: Bellton.

N. biformis, Lx.
N. Clarksoni, Lx.
N. cordata, Br.

Kittanning C.	Kanawha: Blacksburg.
Kittanning Coal
Waynesburg C.	Monongalia: Cassville.
Waynesburg C	Doddridge: W. Union.
Upper Barrens	Monongalia: Brown's
	Mills.

N. dictyopteroides, F. & W.

Waynesburg C.	Doddridge: W. Union.
Upper Barrens.	Marshall: Bellton.

N. fimbriata, Lx.

Waynesburg C. Monongalia: Arnettsville.
L. Productive C. M. Kanawha River.

N. flexuosa, Br.

Waynesburg C.	Doddridge: W. Union.
Upper Barrens.	Monongalia: Brown's
	Mills.
Lower Barrens	Ohio: Wheeling.
L. Productive C	M. Upper Freeport C.
L. Productive C.	M. Kittanning C.

Var. longifolia, F. & W.
N. gibbosa, Lx.

Waynesburg C.	Doddridge: W. Union.
Upper Barrens.	Monongalia: Brown s
	Mills.

N. Grangeri, Br.
N. heterophylla, Br.

Lower Barrens	Ohio: Wheeling.
Upper Barrens.	Monongalia: Brown's
	Mill.s
L. Productive C.	M. Kittanning Coal.

N. hirsuta, Lx.

Upper Barrens.	throughout.
Lower Barrens	Ohio: Wheeling.
L. Productive C	M. Kittanning Coal
L. Productive C.	M. Upper Freeport C

N. Loschei, Br.

Waynesburg C.	Monongalia: Cassville
Upper Barrens.	Monongalia: Brown's
	Mills.
Lower Barrens.	Ohio: Wheeling.

N. obscura, Lx.	Upper Barrens	Monongalia	Brown's
N. odontopteroides, F. & W.	Waynesburg C.	Monongalia: Cassville, Dent's Run
	Waynesburg C.	Doddridge: W. Union
	Upper Barrens.	Monongalia; Brown's Mills
N. plicata, Sternb.	Upper Barrens	Monongalia: Brown's Mills.
N. rarinervis, Bunby.	Lower Barrens,	Ohio: Wheeling.
	L. Productive C. M.	Upper Freeport C.
	L. Productive C. M.	Kittanning Coal.
N. Schenchzerii, Br.	Waynesburg C.	Monongalia: Cassville
	Upper Barrens.	Monongalia: Brown's Mills.
N. Smithii, Lx.	Conglomerate.	Fayette. Quinnimont, Sewell.
N. tenuifolia, Br	Waynesburg C.	Monongalia: Arnettsville.
	L. Productive C. M.	Kanawha: Blacksburg, Coalburg.
	Conglomerate,	Fayette: Quinnimont, Sewell.

ODONTOPTERIS.

O. densifolia, F. & W.	Waynesburg C.	Monongalia: Cassville
O. gracillima, Newby.	Conglomerate,	Fayette: Quinnimont, Sewell.
O. nervosa, F. & W.	Waynesburg C.	Monongalia: Cassville,
	Waynesburg C.	Doddridge: W. Union.
O. neuropteroides, Newby.	Conglomerate.	Fayette: Quinnimont, Sewell.
O. Newberryi, Lx.	Conglomerate.	Fayette: Quinnimont, Sewell.
O. obtusiloba.	Conglomerate.	Fayette: Quinnimont, Sewell.
var. *rarinervis*, F. & W.	Upper Barrens.	Marshall, near Bellton.
O. pachyderma, F. & W.	Waynesburg C.	Monongalia: Cassville.
	Waynesburg C.	Monongalia: Dent's Run.
O. Reichiana, F. & W.	Waynesburg C.	Monongalia: Cassville.
	Upper Barrens	Monongalia: Brown's Mills
O. sphenopteroides, Lx.	L. Productive C. M.	Kanawha: Blacksburg
	L. Productive C. M.	Boone: Short creek.
O. subcuneata, Bunby	Waynesburg C.	Monongalia: Arnettsville.
	L. Productive C. M.	Upper Freeport C.
O. tenuinervis, Lx	Upper Barrens.	Monongalia: Brown's Mills.
?O. Werthenii Lx	Waynesburg C.	Monongalia: Cassville.

PECOPTERIS.

P. angustipinna. F. & W. — Waynesburg C. Monongalia: Cassville.
Waynesburg C. Monongalia: Dent's Run.
Waynesburg C. Doddridge: West Union.

P. arborescens(Schl.). Br. — Waynesburg C. Monongalia: Cassvilie.
Waynesburg C. Doddridge: W. Union.
Upper Barrens. Marshall. Bellton.
L. Productive C. M. Upper Freeport C.

Var. *integrapinna.* F.&W. — Waynesburg C. Monongalia: sta.?
Upper Barrens. in coal. Marshall: Tyler.

P. arguta. Sternb. — Waynesburg C. Monongalia: Cassville.
Waynesburg C. Doddridge: W. Union.

P. asplenoides. F. & W. — Waynesburg C. Monongalia: Cassville.

P. Bucklandii. Br. — Lower Barrens. Ohio: Wheeling.

P. Candolliana. Br. — Lower Barrens. Ohio: Wheeling.
Waynesburg C. Doddridge: W. Union.

P. Cisti. Br — L. Productive C. M. sta.?

P. dentata. Br. — Waynesburg C. Monongalia: Cassville.
Lower Barrens. Ohio: Wheeling.

Var. *crenata.* F & W. — Waynesburg C. Monongalia: Cassville.

Var. *parva.* F.& W. — Waynesburg C. Monongalia: Cassville.

P. elegans. Germ. — Waynesburg C. Monongalia: Cassville.

P. elliptica. Bunbv. — Waynesburg C. Monongalia: Cassville.

P. emarginata(Goep). Bunby — Waynesburg C. Monongalia: Cassville.
Upper Barrens. Marshall: Bellton.

P. Germari(Weiss.). F.& W. — Waynesburg C. Monongalia: Cassville.
Waynesburg C. Doddridge: W. Union.

Var. *crassinervis.* F.& W — Waynesburg C. Doddridge: W. Union.

Var. *cuspidata.* F.& W. — Waynesburg C. Doddridge: W. Union.

P. goniopteroides. F.& W. — Waynesburg C. Monongalia: Cassville.

P. Heeriana. F.& W. — Waynesburg C. Monongalia: Cassville.

P. imbricata. F.& W. — Waynesburg C. Monongalia: Cassville.

P. inclinata. F.& W. — Waynesburg C. Monongalia: Cassville.

P. lanceolata. F.& W. — Waynesburg C. Monongalia: Cassville.
Waynesburg C. Ohio: Moundsville.
Upper Barrens. Marshall: Belton.

P. latifolia. F. & W. — Waynesburg C. Monongalia: Cassville.
Upper Barrens. Marshall: Bellton.

P. longifolia. Br. — Waynesburg C. Monongalia: Cassville.

P. meriuniopteroides. F.& W. — Waynesburg C. Monongalia: Cassville.

P. microphylla. Br. — Waynesburg C. Monongalia: Cassville.

P. Miltonii. Br. — Waynesburg C. Monongalia: Cassville.
Waynesburg C. Marshall: W. Union.
L. Productive C. M. Ohio: Wheeling.
Lower Barrens. Ohio: Wheeling.

P. notata. Lx. —

P. oreopteridia(Schloth.). Br. — Upper Barrens. Marshall: Bellton.

P. oroides. F. & W. — Upper Barrens. Marshall: Bellton.

P. pachypteroides. F. & W. — Waynesburg C. Monongalia. Cassville.

P. pennaeformis. Br — Waynesburg C. Monongalia: Cassville.

Var. *latifolia.* F.& W. — Waynesburg C. Monongalia: Cassville.

P. platynervis. F.& W. — Waynesburg C. Monongalia: Cassville
and Dent's run.

P. pteroides, Br

Waynesburg C Doddridge West Union.
Waynesburg C. Monongalia Cassville
and Arnettsville.
Lower Barrens Ohio Wheeling.
L. Productive C. M. Kanawha Sta ?

P. rarinervis, F.& W.
Waynesburg C Monongalia Cassville.

P. rotundifolia, F.& W.
Waynesburg C Monongalia Cassville.
Upper Barrens. Monongalia Wisee.

P. rotundiloba, F.& W.
Waynesburg C Monongalia Cassville.

P. Schimperana, F.& W.
Waynesburg C. Doddridge W. Union
and Monongalia. Cassville.

P. subfalcata, F.& W.
Waynesburg C. Monongalia. Cassville

P. tenuinervis, F.& W.
Waynesburg C. Monongalia. Cassville
and Arnettsville.
Upper Barrens. Monongaha: south of
Jollytown, Pa., in W. Va.

P. villosa, Br.
L. Productive C. M. Upper Freeport C
L. Productive C. M. Kittanning Coal.

PSEUDOPECOPTERIS (from Pecopteris.).

P. muricata (Br.), Lx.
Conglomerate. Fayette: Quinnimont,
Sewell.

P. nervosa (Br.), Lx.
Conglomerate. Fayette: Quinnimont,
Sewell.

P. Pluckenetii (Br.), Lx.
Lower Barrens. Ohio: Wheeling.

P. latifolia (Br.), Lx.
L. Productive C.M. Kanawha: Coalmont.

P. macilenta (Llott. . Lx.
L. productive C. M. Kanawha: Coal
mont.
Conglomerate. Fayette. Quinnimont
and Sewell.

P. cordato-ovata Weiss. . Lx.
Waynesburg C. Monongalia: Cassville.

P. obtusiloba Br.). Lx.
L. Productive C. M. Kanawha Wy
oming Mine near Upper Creek.
Conglomerate. Fayette: Quinnimont
and Sewell
L. Productive C. M. Boone: Peytona.

P. obtusiloba, var. *dilatata*
(Ll. & H.). Lx. L. Productive C. M. Boone: Short Creek.

P. Virginiana Meek). Lx.
Pocono or Vespertine. Greenbrier:
Lewis Tunnel.

P. spinulosa, Lx.
Lower Barrens. Ohio Wheeling

P. Andreana (Roehl.). Lx.
L. Productive C. M. Kanawha. Blacks-
burg, Coalmont

RHABDOCARPUS.

R. oblongatus, F. & W.
Waynesburg C Monongalia: Cassville.

R. Boeckschianus, G. & B.
L. Productive C M Kanawha: Campbell's Creek.

R. clavatus, (St.). Gein.
L. Productive C. M. Kanawha: Blacks-
burg.
L. Productive C. M. Kanawha: Malden
Mines.

R. multistriatus(Presl.). Lx. L. Productive C. M. Kanawha. Campbell's Creek.

R. tenax, Lx. L. Productive C. M. Kanawha: Campbell s Creek.

RHACOPHYLLUM.

R. filiciforme Guth.). Schp. Lower Barrens. Ohio: Wheeling.

Var. *majus*. F. & W. Waynesburg C. Monongalia: Cassville.

R. laciniatum. F. & W. Waynesburg C. Monongalia: Cassville.

R. lactuca St.). Schp. Waynesburg C. Monongalia: Cassville.

Waynesburg C. Doddridge: West Union.

R. speciosissimum. Schp. Waynesburg C. Doddridge: West Union.

SAPORTEA.

S. grandifolia. F. & W. Waynesburg C. Monongalia: Cassville.

S. salisburinides. F. & W. Waynesburg C. Monongalia: Cassville.

SIGILLARIA.

S. approximata. F. & W. Waynesburg C. Monongalia: Arnettsville.

S. Menardii. Br. Upper beds above the Pittsburg coal.

S. scutellata. Br. L. Productive C. M. Sta. ?

SPHENOPHYLLUM.

S. angustifolium. Germ. Waynesburg C. Monongalia: Cassville, Dent's Run.

Upper Barrens. Monongalia: Wadestown.

S. densifoliatum. F. & W. Waynesburg C. Monongalia: Cassville.

S. erosum, L. & H. (trifoliatum) Lower Barrens. Ohio: Wheeling.

S. filiculme, Lx. Lower Barrens. Ohio: Wheeling.

Waynesburg C. Monongalia: Cassville.

Waynesburg C. Doddridge: W. Union.

S. latifolium. F. & W. Waynesburg C. Doddridge: W. Union.

Waynesburg C. Monongalia: Cassville.

S. longifolium. Germ. Waynesburg C. Monongalia: Cassville.

Waynesburg C. Doddridge: W. Union.

S. oblongifolium. Germ. Waynesburg C. Monongalia. Cassville.

Upper Barrens. Monongalia: south of Jollytown, Pa.

S. Schlotheimii, Br. L. Productive C. M. Upper Freeport C.

L. Productive C. M. Kittanning Coal.

S. tenuifolium. F. & W. Waynesburg C. Monongalia: Cassville.

Upper Barrens. Monongalia: south of Jollytown, Pa.

Waynesburg C. Doddridge: W. Union.

S. tenuifolium and *S. longifolium* are considered by Lesquereux to belong to *S. angustifolium*, Germ.

SPHENOPTERIS.

S. acrocarpa, F. & W. Waynesburg C. Monongalia: Cassville.

S. adiantoides. L. & H. Conglomerate. Fayette: Quinnimont, Sewell.

S. auriculata. F. & W. Waynesburg C. Monongalia: Cassville.

S. coriacea, F. & W.	Upper Barrens. Monongalia: Brown s Bridge.
S. dentata, F. & W.	Waynesburg C. Monongalia: Cassville.
S. Dubuissionis, Br.	Lower Productive C.M. Kanawha Co.
S. foliosa, F. & W.	Waynesburg C. Monongalia: Cassville. Upper Barrens. Monongalia: South of Jollytown. Pa.
S. tenella, Br.	Lower Productive C.M. Kanawha: Malden Mines
S. elegans, Br.	Lower Productive C.M. Kanawha: Malden Mine
S. furcata, Br.	Lower Barrens. Ohio: Wheeling.
S. hastata, F. & W.	Waynesburg C. Monongalia: Cassville
S. Hildrethii, Lx.	L. Productive C.M. Kanawha Co.
S. Heninghausii, Br.	Conglomerate. Fayette: Quinnimont Sewell.
S. Lescuriana, F. & W.	Waynesburg C. Doddridge: West Union.
S. minutisecta, F. & W.	Waynesburg C. Doddridge: West Union. Lower Barrens. Ohio: Wheeling.
S. pachyueris, F & W.	Waynesburg C. Doddrige: West Union.
S. tridactyllites, Br.	L. Productive C.M. Kanawha Co.
S. Grovenhorstii, Br.	L. Productive C. M. Kanawha Blacksburg.

SYRINGODENDRON.

S. pes-capreoli,* St.	Lower Barrens. Ohio: Wheeling.

T.ENIOPTERIS.

T. Lescuriana, F. & W.	Waynesburg C. Monongalia: Cassville.
T. Newberryana, F. & W.	Waynesburg C. Monongalia: Cassville.?
Var. *angusta*, F. & W.	Waynesburg C. Monongalia: Cassville.?

TRIGONOCARPUS.

T. Noeggerathii, Br	L. Productive C. M. Kanawha: Campbell's Creek and Malden Mine.
T. ampullaformis, Lx.	L. Productive C. M. Kanawha: Malden Mine.

TRIPHYLLOPTERIS.

T. Lescuriana(Meek.), Lx.	Vespertine. Greenbrier: Lewis Tunnel.

*Syringodendron pes-capreoli, is Sternberg's genus and species, and may be credited accordingly. Geinitz described it as *Sigillaria Brongniarti*, and transferred Stanberg's sp. to *Sigillaria*. Schimper made Sternberg's species a synonym of Geinitz' *Sigillaria Brongniarti* and Lesquereux transfers the united species to *Syringodendron*, like Schimper, using Geinitz' specific name, which if followed, makes it *Syringodendron Brongniarti*(Gein. Lx. or if Sternberg's specific name is adopted, *Syringodendron pes-capreoli*, St. Either is admissible. I. H. wed Prof Lesqx though Sternberg has priority of name, but probably lacked in satisfactory description, giving Geinitz' name, given above, the preference.—L. D. Lacoe.

INDEX.

ERRATA.

P. 315 line 32 for "or name initial" read: name or initial.
P. 317 line 1 erase "more."
P. 320 line 18, for "alt. 200 ft." read: alt. 2000 ft.
P. 325 line 13. for "Z. apiifolia" read: X. apiifolia.
P. 327 line 13. NECKERIA, as a genus also appears on page 49 Musci. I have not the opportunity to properly determine which should stand: doubtless the moss.
P. 333 line 16 insert:

V. canina, L. var. Muhlenbergii(Torr.). Gray.
Fayette: at foot of cliff near Nuttaliburg—L. W. N.

P. 349 line 24. for "woons" read: woods.
P. 468 line 18. for PHLARIS, read: PHALARIS.

www.ingramcontent.com/pod-product-compliance
Lightning Source LLC
Chambersburg PA
CBHW021657210326
41599CB00013B/1454